WAR PLAY

WAR PLAY

VIDEO GAMES AND THE FUTURE OF ARMED CONFLICT

Corey Mead

AN EAMON DOLAN BOOK
Houghton Mifflin Harcourt
BOSTON NEW YORK
2013

www.hmhbooks.com

Library of Congress Cataloging-in-Publication Data is available.
ISBN 978-0-544-03156-2

Printed in the United States of America
DOC 10 9 8 7 6 5 4 3 2 1

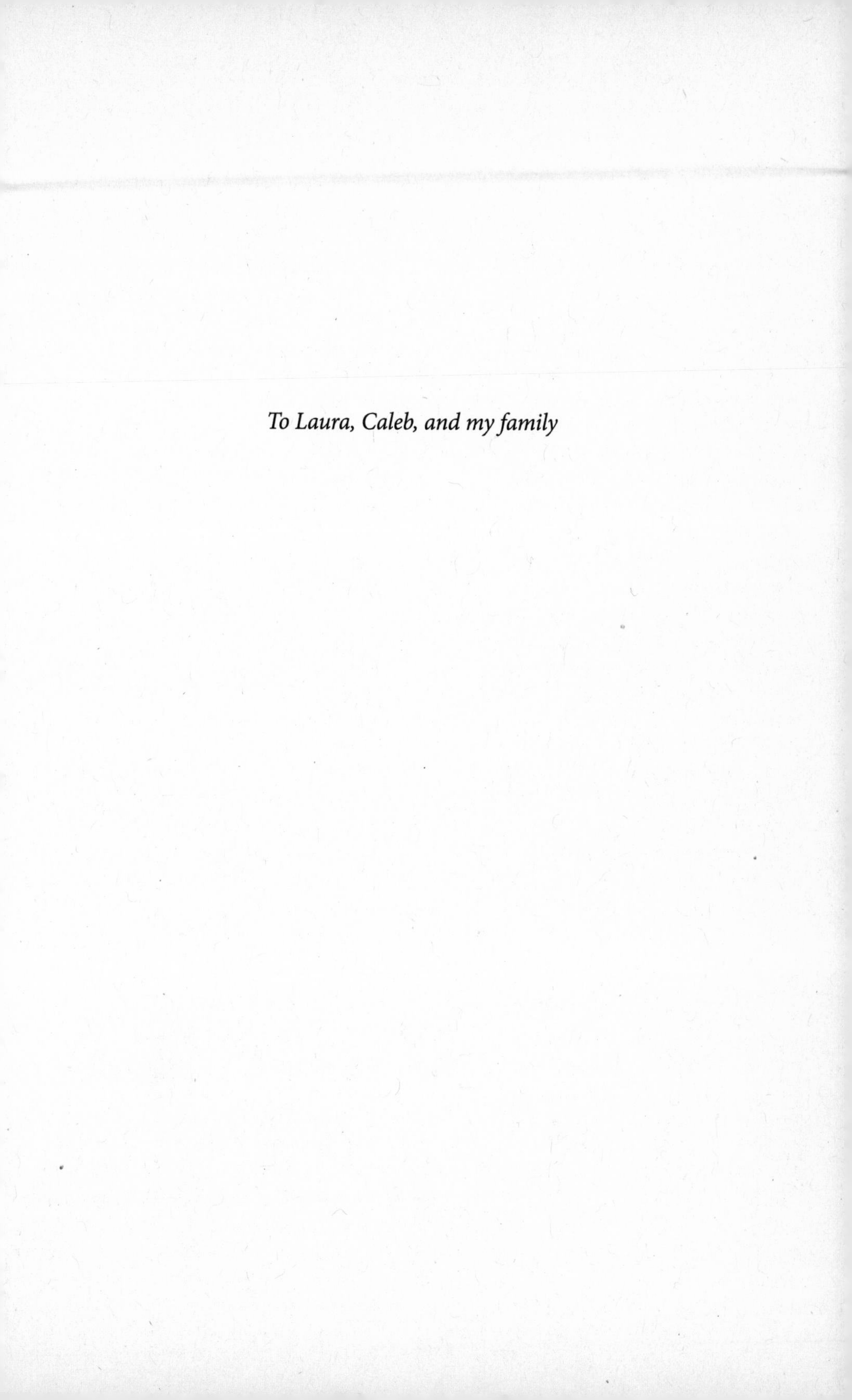

To Laura, Caleb, and my family

CONTENTS

WAR PLAY

INTRODUCTION

ONE JUNE DAY, I visited a brick-and-concrete warehouse on a dead-end street of squat buildings in Playa Vista, California. The outer part of the warehouse housed a suite of glass-fronted offices, while the main room was strewn with sandbags, corrugated metal, piles of fake rubble, and twisted rebar. Placed throughout this room were strategic groupings of "digital flats," large rear-projection screens that employ digital graphics to depict particular settings and geographic locations. The warehouse, a former television studio, was owned by the Institute for Creative Technologies (ICT), a joint venture between the military and the University of Southern California. Funded by the army to the tune of hundreds of millions of dollars, the ICT's declared mission is "to build a partnership among the entertainment industry, Army, and academia with the goal of creating synthetic experiences so compelling that participants react as if they are real." As the ICT's executive director, Dr. Randall Hill, said to me, "One way of seeing our mission, one way we view it, is that we're trying to forge leaders and revolutionize learning — in general, not just in the military. It's really about how you use digital interactive media, forms of media, to aid the learning process."

The main room of the warehouse in which I stood contained *Flat-World,* one of the ICT's earliest projects, described as "a mixed reality environment where users interact with both the physical and virtual worlds seamlessly." *FlatWorld* was conceived of and designed by video game designers, special effects artists, research scientists, and Pentagon personnel working together to create the army version of *Star Trek*'s fictional "holodeck," a simulated-reality facility that mimics the environments of alien planets. The military's goal in creating this type of fully immersive domain was to give soldiers the most accurate training environment possible outside of live field exercises.

As I walked through *FlatWorld,* Jarrell Pair, my guide, led me through a door into a tiny room fronted by a large digital screen. Onto the screen (the "flat") was projected the computer-animated version of a deserted city street lined with squat gray residential buildings, a white mosque with two minarets, telephone wire, and palm trees. A Middle Eastern carpet covered the floor of the room, with pieces of concrete and wrecked furniture heaped in one corner. Broken ceiling panels hung overhead. After instructing me to put on a pair of polarized 3-D glasses, Pair began pressing buttons on a small controller pad. Suddenly, in the open wooden doorway to my right, there appeared a life-size, computer-generated army officer yelling at me that the enemy was approaching. Just then a computer-animated helicopter roared in overhead and began strafing the street, as insurgents and U.S. soldiers appeared along the road, each group firing at the other. One insurgent popped up in the open doorway where the American officer had been. He pointed his machine gun in my direction and started firing, and the wall to my left began sending out virtual clouds of plaster dust, which cleared to reveal pockmarks where the bullets had lodged. The ground in the room began to shake, and the volume of the helicopter overhead and the gunfire in the street increased to the point of near discomfort. A little boy ran into the street and shouted, "U.S.A.! Over here!" Pair pressed another button and a tank rounded the corner at the end of the street, then headed straight toward me. As it bore down, the combination of the noise, the rumbling ground, and the tank cannon pointing at my face stirred genuine anxiety in me. The anxiety built for several seconds until, at the moment when the tank appeared about to run me over, Pair pressed another series of buttons and the room returned to its original state — no tank, no insurgents, no U.S. soldiers, no helicopter, no noise, no rumbling ground, just a panel projecting the digital image of a now empty city street.

With its virtual, immersive nature, *FlatWorld* — currently in use as the Joint Fires and Effects Training System at Fort Sill, Oklahoma — is a large-scale example of the U.S. military's changing approach to training and educating its soldiers. While live field exercises and training manuals are still crucial, they are increasingly being supplemented and supplanted by video games and digital simulations,* which are used to teach everything from battlefield operations to cultural interaction to language skills to weapons handling. Though the specifics vary, today every armed forces service member engages in some form of virtual learning. Helping troops protect themselves or gain the advantage against the endlessly mutating insurgencies that mark today's wars requires a constant shifting of strategies and tactics and the kinds of rapid adjustments in scenarios that print-based manuals, which are updated every six months at the most, can't keep up with. Video games, in contrast, allow for near-instantaneous user modification, meaning that soldiers in the field can, on a daily basis, input the enemy's latest fighting tactics, so that troops who are stateside can keep their training up-to-date. As one Marine officer said to me, events in military gaming are moving "at the speed of war." The military's use of video games also extends beyond the battlefield: games are used to treat soldiers suffering from post-traumatic stress disorder, and they aid veterans who are reintegrating into civil society.

The military's desire to harness game technologies stems in part from the realization that its traditional approach to learning, and to the role of soldiers, often no longer applies. Standardization and functionality, the longstanding military paradigms for instruction, don't

* For simplicity's sake, in this book I use the term *video game* to refer to a variety of interactive digital and virtual applications.

always fit the problems of America's new hybrid wars. The reality is that soldiers are themselves now a form of information technology, responsible for a far broader range of roles, decisions, and systems-based interactions than in any previous conflict. This is the extension of a process that began during World War II, when the military emphasis on selection, classification, and "human factors" training came to the fore.* At the time, military psychologists argued that the "human-machine system," not just the machine itself, was the fundamental military unit, a claim supported by many military leaders. In the intervening years, this view has only become more prominent. Today soldiers' skills are measured largely in relation to the technological systems the soldiers will be using.

To a sizable extent, the military is turning to gaming for scenarios that involve new and unexpected roles for soldiers as well as the mental and physical side effects of multiple deployments. Take the rise of nontraditional soldier roles: soldiers today use games to learn skills such as cultural negotiation, because in our post-9/11 wars they must deal with disputes between tribal elders or with the complexities of building a police force. In past wars, when issues related to civil affairs arose, the government farmed them out to other agencies, but those duties are now increasingly under the Pentagon's control. This, then, is part of why the military now relies so heavily on gaming: it helps to plug the holes, to address the issues that previous military instruction wasn't set up to address.

The hype over video games extends far beyond the military, of course. These games suffuse our popular culture and the lives of our young people, generating more yearly profits than the movie and music industries combined. And though video games have long been criticized as being harmful to kids, even mainstream educators and administrators across the country are beginning to follow the military's lead and treat games as potentially revolutionary educational

* "Human factors" centers on the idea that the effective running of complex technological systems depends as much on the people running them as on the equipment itself.

tools. Edward O. Wilson, professor emeritus at Harvard, caused a stir by declaring, "Games are the future of learning." President Obama, meanwhile, has identified the creation of good educational software as one of the "grand challenges for American innovation." (To meet this challenge, the Obama administration created the Advanced Research Projects Agency for Education, which has as a major goal the creation of educational software "as compelling as the best video game.") For many years, a rapidly expanding "serious games" movement has been pushing the use of video games as teaching tools in schools and workplaces.

As we will see, if games are the future of learning, it is a future that the military already inhabits. At the same time, the military's use of video games is just the latest entry in its long history of learning innovation. Though the fact is rarely recognized, over the past century the military has helped shape the contours of American education. Mass standardized tests? Computer-based learning? Adult education? A functional approach to education? All of these have been launched or refined by the military. We don't have to look far to see that the paradigms of military training—standardization, efficiency, functionality—are everywhere apparent in our schools and workplaces alike. If the historical pattern continues to apply, we can expect to see the military's use of video games having a deep influence on our public institutions, not only in terms of methods of instruction but in regard to the skills that people will be expected to master. And yet, while the growth of virtual and game-based learning represents a potential sea change in American education, the military's role in this potential transformation has gone almost entirely unnoticed by both the public and the media.

What is unique about the military's employment of video games is that it is deploying them on a broad, institution-wide scale. While any number of other institutions—businesses, schools, health-care organizations, government agencies—are using games for learning, the military is the first organization that has *substantially* moved into video games, using them at every organizational level for a broad array of purposes. Moreover, although commercial interests have had by far the largest influence on video games as a cultural phenomenon, the military is the first institution to use video games in direct support of state purposes, and the most serious of purposes at that: the use of force to protect the state's interests.

An overarching theme in this book is the military's long history of technical and instructional innovation. We see that today in its use of video games, but the scenario has shifted so that the military is now following the entertainment industry, not leading it. At the same time, the military — that conservative, hidebound institution — is showing more flexibility and openness about learning than our public schools are. The Pentagon may be following the entertainment industry, but it is leading the education industry. This, too, has gone entirely unremarked upon in the media and public spheres.

One of the ironies here is that gaming technologies were initially created for military needs and were developed in military-sponsored projects and labs. For decades the military took the lead in financing, sponsoring, and inventing the technology used in video games, while game companies were the happy beneficiaries. The entire game industry rests on a technological foundation established by large amounts of military-funded research and infrastructure, including advanced computing systems, computer graphics, and the Internet.

Not until the late 1990s did the technological balance between the military and the game industry shift for good, as the military's budget was reduced and as cheaper, smaller, more powerful computers became commercially available. Today the video game industry far surpasses the military in technological expertise, with the result that the military now procures its game technology from commercial game companies. By partnering with these companies, it is granted access to proprietary technology, while game makers receive the military's money and occasionally its official stamp of approval. This exchange has led scholars to dub the partnership between the military and the video game industry the "military-entertainment complex."

While the army is the largest military user of video games, the other services rely on gaming technology—and, more broadly, on modeling and simulation—as well. According to the Defense Intelligence Agency, more than three hundred virtual worlds are in development for military purposes, and that figure is likely to grow. The 2012 Department of Defense budget allocated at least $224 million specifically for modeling and simulation; the research firm Frost & Sullivan predicts that DoD spending on modeling and simulation will reach $24.1 billion by 2015.

Two questions that remain unanswerable are the number of games and simulations that service members will encounter in their training and what percentage of their training will be virtual. The simple fact is that nobody knows, because no larger set of requirements is guiding virtual training. It's up to each company commander to determine how much simulation he or she wants to use, but that's not something the military keeps track of. Thus, the only concrete answer that my interviewees (backed up by secondary sources) have given me is that every service member encounters game-based learning at some point in his or her training.

There are major consequences of the military's video game use both within and beyond the services. For the military itself, the combined outcome is potentially paradigm-shifting. As defense expert Peter Singer points out, "game-based training can be tailored to specific scenarios as well as to an individual's own rate of learning, sped up or slowed down based on how quickly he or she is learning." This is not to mention the untold millions of dollars in savings that virtual training enables, a factor that will only grow in importance as defense budgets shrink.

Taken as a whole, the trend lines for gaming in the military point to enormous growth. While the army has been ahead of the other services in exploiting the benefits of game-based learning, for example, this will not be the case in a few more years. At the same time, the Pentagon, as I mentioned, has yet to develop an overall policy or set of metrics for its use of video games, even as its reliance on games continues to grow. There remain a number of critical questions that the Pentagon must answer, including how game-based learning can be placed in its proper perspective and what an effective balance between virtual training and "muddy boots" training might be.

These concerns become even more pressing in regard to such issues as post-traumatic stress disorder. The use of video games to treat PTSD, for example, points to larger issues involving the soaring rate of mental health problems, as well as suicide, in the military.* This is an issue that the entire country, not just the Pentagon, should be debating, and yet it remains distressingly absent from our public discourse.

Nor do the concerns end there. As defense expert John Arquilla points out, issues surrounding virtual reality become particularly blurred in the area of cyberwarfare, which to date has been almost exclusively a virtual phenomenon. Military thinkers warn that in the future cyberwarfare will increasingly intrude on the real world, as developments like the Stuxnet and Flame computer viruses deployed by the United States and Israel against Iran's nuclear program illustrate, in Arquilla's words, "how zeros and ones can have real effects on physical phenomena."

For the past several years I have been investigating the historical foundations of the military-entertainment complex as well as how it operates today. During this time I have traveled the country interviewing the leaders who are pioneering the military's use of video games as learning tools. I have visited the sites where the military's games are being developed and put into practice. I have spoken with dozens of military, civilian, and academic experts. This book is an account of what I've learned during my travels and my research. Readers looking for the technical details of military video games will not find them

* One soldier with PTSD that I profile in this book had to fight the VA for two full years to get his benefits, and those were granted only after he wrote a letter of appeal to President Obama. Many other vets with PTSD whom I interviewed spoke about the same thing — the incredible hassle of cutting through the VA's red tape.

here; nor will readers looking for a detailed analysis of commercial first-person shooters like the *Call of Duty* series find what they are seeking. But anyone concerned with the present and future of war and education will find in the following pages a map of the immense—and consequential—changes swirling around both.

CHAPTER 1

The Rise of the Military-Entertainment Complex

THE ORIGINS OF THE U.S. military's involvement with video games lie in its century-old status as this country's primary sponsor of new technologies. A quick checklist of the technologies that either stem from or were significantly refined in defense-funded contexts shows how pervasive the military's influence has been: digital computers, nuclear power, high-speed integrated circuits, the first version of the Internet, semiconductors, radar, sonar, jet engines, portable phones, transistors, microwave ovens, GPS — the list goes on. As Ed Halter writes in his book *From Sun Tzu to Xbox*, "The technologies that shape our culture have always been pushed forward by war."

Take the creation of the key technological innovation of the past several decades: the digital computer. Specifically, the combination of military-sponsored technological advancements and military-related strategic and tactical needs during World War II led directly to the computer's invention. While private industry may eventually have developed what we now think of as the digital computer, the huge amount of research funding provided by the military, coupled with a desperate need to win the war, pushed this development up by years, if not decades. The military remained the key influence on the advancement of digital computers well into the 1960s.

Early in the twentieth century, the rising prominence of artillery made the calculation of ballistics a tactical necessity. Up to and throughout World War I, ballistics data had primarily been tabulated by hand, even as the growing number and variety of modern weapons called for ever more complex calculations, which became the province of skilled mathematicians known as "computers." By World War II, however, advances in air and weapons systems required quicker means of calculation. To meet this requirement, the military sponsored the creation of the Electronic Numerical Integrator and Computer (ENIAC), popularly regarded as the world's first digital computer. Though ENIAC wasn't completed until the fall of 1945, after the war had drawn to a close, the military connection remained strong: ENIAC's first task was to provide calculations used to plan the detonation of the hydrogen bomb.

Military-sponsored technological innovation continued apace for the next two decades, as the Department of Defense and its subagencies underwrote the great majority of computer and electronics research and development. In the years after the war's end, the DoD founded a number of grant-giving agencies that continue to underwrite new technology today. Among these agencies were the highly influential Advanced Research Projects Agency (ARPA), now known as the Defense Advanced Research Projects Agency (DARPA); the Army Office of Scientific Research, now the Army Research Laboratory; and the Office of Naval Research. Throughout the 1950s and '60s, the military remained what historian Paul Edwards notes was "the proving ground for initial concepts and prototype machines."

Paralleling the rise of these DoD-operated institutions was an increasingly large defense contracting sector, ranging from companies whose sole focus was military contracting to larger, more diverse corporations, such as IBM, Raytheon, and General Electric, whose success was built on a combination of military subsidies and commercial sales. Beginning in the 1960s, the private electronics sector also experienced unprecedented growth, compelling it to begin pouring money into its own research and development. Despite this sector's self-financed efforts, however, military funding continued to be the primary force spurring the creation of new technologies.

There were several reasons for the military's intense interest in and financing of computers, but none was as important as the huge information-processing needs of what had become an immense bureaucracy. With the advent of the Cold War, American supremacy was thought to hinge on the maintenance of a robust military, whose efficient functioning required number-crunching on a vast scale. Computers greatly accelerated this process, further sparking the growth and increasing complexity of military bureaucracy.

Beyond the military's bureaucratic needs, the drive for computerization reflected a broader ideological shift. In the late 1940s and '50s, computers were enlisted as tools for the newly popular practice of tackling society's biggest problems with seemingly objective statistical and mathematical tools. In fact, computers were for many years developed specifically to meet this function. Even as the private computer industry continued to expand during the early 1960s, the military, and the defense industry as a whole, remained the prime buyer and sponsor of computer-related technology.

The other major beneficiary of computer-oriented military funding during this period was academia; the Pentagon and ARPA underwrote research in the field at such prestigious institutions as Harvard, Johns Hopkins, Stanford, and UCLA. Perhaps most notably, the Massachusetts Institute of Technology, along with its groundbreaking artificial intelligence program, received the majority of its computer-related research money from the military. In their superb analysis of the video game industry, *Digital Play*, Stephen Kline, Nick Dyer-Witheford, and Greig de Peuter draw on this exchange to note that the "military-industrial-academic complex provided the triangular base from which the information age would be launched."

Spacewar! and Beyond

The roots of the military's historical involvement with video games extend beyond its sponsorship of computers. For several decades — from the 1960s to the early 1990s — the armed forces took the lead in financing, sponsoring, and inventing the specific technology used in video games. Without the largesse of such military agencies as DARPA, the technological foundation on which the commercial game industry rests would not exist. Advanced computing systems, computer graphics, the Internet, multiplayer networked systems, the 3-D navigation of virtual environments — all these were funded by the Department of Defense.

Virtual military training dates back to the late 1920s, when Edwin Link, the son of an organ and automatic piano maker, developed the first flight simulator, which was made of wood and powered by organ bellows. Video games, however, derive from preparations for nuclear war and space exploration; arguably the first digital game, a faux-military simulation, was in fact called *Spacewar!* The game was invented in 1962 by twenty-three-year-old Steve Russell and his cohorts in the fictitious Hingham Institute Study Group on Space Warfare, a collection of like-minded, Pentagon-funded engineering graduate students at MIT. Russell and his friends were as fascinated by science fiction as they were by their basement lab's latest acquisition: a Programmed Data Processor-1, or PDP-1, one of the first microcomputers, which Russell describes as "the size of about three refrigerators," with "an old-fashioned computer console" and "a whole bunch of switches and lights."

The PDP-1's manufacturer had shipped the computer to MIT in the hope that the electrical engineering department could put it toward some new and intriguing use, though building the world's first video game could hardly have been what the manufacturer had in mind. For a time the PDP-1 just sat idle in the corner of the engineering lab. Russell was "itching to get his fingers" on the new machine, however, so he and his friends began discussing what they could do with this new mini computer. According to Russell, "Space was very hot at the time — it was just when satellites were getting up and we were talking about putting a man on the moon. So we said, gee, space is fun, and most people don't appreciate how to maneuver things in space. So I wrote a demo program that had two spaceships that were controlled by the switches on the computer."

Russell's main influence in programming *Spacewar!* was Edward "Doc" Smith's science-fiction "space opera" *Lensman,* which appeared serially in magazines before being reworked into highly successful books. Russell and his MIT coworkers were big fans of *Lensman.* "The details were very good and it had an excellent pace," Russell says. "[Smith's] heroes had a strong tendency to get pursued by the villain across the galaxy and have to invent their way out of their problem while they were being pursued. That sort of action was the thing that suggested *Spacewar!* He had some very glowing descriptions of spaceship encounters and space fleet maneuvers." "Glowing" is certainly an accurate description, as is evident in this sample from one of the *Lensman* books:

> Beams, rods, and lances of energy flamed and flared; planes and pencils cut, slashed, and stabbed; defensive screens glowed redly or flashed suddenly into intensely brilliant, coruscating incandescence. Crimson opacity struggled sullenly against violet curtains of annihilation. Material projectiles and torpedoes were launched under full-beam control; only to be exploded in mid-space, to be blasted into nothingness or to disappear innocuously against impenetrable polycyclic screens.

Russell felt that by "picking a world which people weren't familiar with" — that is, space — "we could alter a number of parameters of the world in the interests of making a good game and of making it possible to get it onto a computer." In the game, two players used switches and knobs to maneuver spaceships through the gravity field of a star while firing missiles at each other. The fuel and the missiles were limited, as they would be in real life; adding to the pressure, players also had to avoid colliding with the star as they fired their weapons. Players could launch their ships into hyperspace at the last minute to avoid incoming missiles, but the ships would reenter the game at random locations, with each reentry increasing the chances that the craft would explode. Graphics-wise, *Spacewar!* was quite primitive: the spaceships were little more than green blips on the murky blue-black background of the oscilloscope screen. Irritated by the inaccuracy of the game's initial star field, one of Russell's coworkers eventually rewrote the script based on real star charts.

Spacewar!'s originality derived from the interface of the PDP-1, which came equipped with a keyboard and a circular monitor. As Kline, Dyer-Witheford, and de Peuter write, the game's "radical innovation" was that it featured "interface controls for navigation and made the screen a graphic input to the player." These twin features of navigation and display are, the authors note, "the foundation of digital interactive entertainment — the crucial 'core design' subsequent hardware and software designers would work up and sophisticate through generations of games." Russell himself had wondered whether there might be a way to commercialize the game in order to make a profit from it. After a week's contemplation, however, he decided that no one would be willing to pay money for it. Instead, he and his friends just gave the source code to anyone who asked for it.

Spacewar! was an immediate hit among the growing network of computer programmers occupying university research institutes nationwide. Within a year, the game had grown so popular that Stanford University's Computer Studies Department had to initiate a "no *Spacewar!* during business hours" policy. By the mid-1960s, a copy of the game was on virtually every research computer in the country, whether in academia, industry, or the military.

Russell and his MIT associates were enthusiastic members of the emerging subset of computer virtuosos known as "hackers" — those who experimented with the programming capabilities of computers for the sheer fun of it. Young, male, and white, both nerdy and counterculture-cool, these hackers were subsidized by the burgeoning military-industrial complex, with their research going to fight the Cold War. The shock of the Soviets' 1957 Sputnik launch had led to vastly increased funding for science and technology, most of it channeled through the Pentagon's Advanced Research Projects Agency. Nuclear mobilization, ballistics, missilery, space defense—these were the concerns of the Pentagon and of policymakers alike. Hackers such as Russell and his friends occupied a precarious position in this new environment. They took their money and much of their guidance from the military, and yet their ethos was one of freedom and playful exploration, and they were harshly disillusioned by Vietnam and, later on, by Watergate.

It would be unfair to say that the hackers' playful spirit was at odds with their military mandate, however. In fact, experimentation and whimsy were *encouraged* in places like the MIT computer engineering lab as a way of expanding the heretofore limited applications of computers. Until the early 1960s, computers were envisioned solely as sophisticated calculators and machines for modeling. Russell and other young hackers introduced the radical notion that computers could be tools not only for calculations but also for entertainment. *Spacewar!* wasn't exciting because of its technology; it was exciting because it introduced a whole new way of thinking about computers—namely, that they could be sources of pleasure. Within a few years, this emphasis on enjoyment became the heart of the growing video game industry. So even though, as Ed Halter writes, video games "were not created directly for military purposes," they nonetheless "arose out of an intellectual environment whose existence was entirely predicated on defense research."

The military's specific interest in computer-based war gaming can be traced to the late 1970s, when the Army War College introduced the board game *Mech War* into its staff officer training curriculum. Much more common during this period, however, was the development of high-end computer simulations, not games, for military training. In the 1980s, collaborators from the military, the entertainment industry, and academia began building "distributed interactive simulations" (DIS) — simulations that use distributed software or hardware to create virtual theaters of war, in which participants could interact in real time. These simulations employed the latest advances in computer graphics and virtual-reality technology, which added to the immersive qualities of their synthetic environments. As DIS technology continued to evolve into the next decade, an increasing focus on content and on compelling narratives brought these simulations closer in basic form to commercial video games.

The military's interest in the kinds of video games popular today dates to 1980, when Atari released its groundbreaking *Battlezone*. Not only did *Battlezone* evoke a three-dimensional world, as opposed to the two-dimensional worlds of such previous arcade hits as *Asteroids* and *Tempest*, but players viewed the action from a first-person perspective, as if they themselves were tank gunners peering through their periscopes at the battlefield outside — in this case, a spare moonscape with mountains and an erupting volcano in the distance. This first-person element made *Battlezone* a direct ancestor of today's enormously popular first-person shooters.

Soon after *Battlezone* took off, the army's Training and Doctrine Command (TRADOC) requested Atari's help in building a modified version of the game that could be used as a training device for the then-new Bradley infantry fighting vehicle. General Donn Starry, the head of TRADOC at the time, had recognized early on that soldiers would be more responsive to electronic training methods than to print- and lecture-based ones. "[Today's soldiers have] learned to learn in a different world," Starry told a TRADOC commanders' conference in 1981, "a world of television, electronic toys and games, computers, and a host of other electronic devices. They belong to a TV and technology generation . . . [so] how is it that our soldiers are still sitting in classrooms, still listening to lectures, still depending on books and other paper reading materials, when possibly new and better methods have been available for many years?" Yet while *Army Battlezone* (also known as *Bradley Trainer*) was eventually produced, the game was never used to train any actual soldiers.

The military's digital efforts took a major step forward with DARPA's construction of *SIMNET,* a real-time distributed networking project for combat simulation. Until the 1980s, simulators had been built as stand-alone systems that focused on such specific tasks as piloting a tank and landing a jet on an aircraft carrier. Each of these systems cost tens of millions of dollars — often twice the amount of the real systems for which the soldiers were training. To rectify this expensive and unwieldy practice, in 1982 DARPA drafted the help of air force captain Jack A. Thorpe, who years earlier had floated the idea that simulators did not need to physically replicate the full vehicles they were representing but could simply be used to enhance the training for these vehicles. Take aircraft: there was no need to use simulators to teach an air force pilot everything he needed to know about flying; simulators could train him only in things that he couldn't learn from flying during peacetime. Why not, Thorpe asked, determine first which training functions were needed and then base the simulator hardware on that?

Thorpe's experience with simulators began in 1976, when he worked as a research scientist in flight training at Williams Air Force Base in Arizona. Tasked with improving the state of flight simulators, which at the time were three-story mechanical contraptions in which pilots were shaken around like leaves, he looked for a way to change these single-pilot machines into ones that could teach group skills. "Group interactions are the most complicated combat operations," he says. "They also tend to be the ones in which the costs of screwing up are the highest. Yet because it is so difficult and expensive to organize groups, pilots get very little training in collective skills. They have to learn these skills on the job, during combat, which makes casualties disproportionately high during the first few missions."

To rectify the situation, Thorpe conceived of a network — anything from dozens to hundreds of individual simulators all interacting with each other. He thought it was wasteful for simulator training devices to focus on individual service members; the network he envisioned would allow for a collective training experience centered on entire crews and units.

By the time of Thorpe's DARPA appointment in the early 1980s, the environment seemed ripe for him finally to put his networking concept into practice. ARPANET — the forerunner of the Internet — had exploded onto the military scene and was generating a great deal of positive interest in the science of networking. Aware that building the kind of system he envisioned would be economically infeasible, Thorpe looked to affordable, non-DoD technology such as computer and video games to make his vision a reality. He hired military contractor Bolt, Beranek and Newman to develop the networking and system software necessary to bring *SIMNET* — that is, simulator networking — to life. The originality of Thorpe's vision later prompted *Wired* magazine to declare, "William Gibson didn't invent cyberspace, Air Force Captain Jack Thorpe did."

By January 1990, the first *SIMNET* units were finally ready to go. The army stepped in first, buying several hundred units for its Close Combat Tactical Trainer system. *SIMNET*'s training value became apparent one year later, during the first Gulf War. In the war's most significant engagement, known as the Battle of 73 Easting, the U.S. 2nd Armored Cavalry Regiment destroyed dozens of Iraqi fighting vehicles in just under two hours, while killing or wounding more than six hundred Iraqi soldiers. Because the 2nd Armored Cavalry had prepared for the war by training extensively on *SIMNET*, the military decided to use the Battle of 73 Easting as a model for future networked training. The goal was to provide a much more rounded experience of battle than simulation had previously allowed for, one that emphasized the stresses and fears, the emotional experience of war, as much as it did the tactical ones. To this end, the *SIMNET* team assembled reams of data on 73 Easting: extensive interviews with 150 participants, radio and tape recordings from the battle, overhead photographs of the skirmishing, action logs, even a step-by-step re-creation on the actual battlefield by soldiers from the 2nd Cavalry. The results of this effort pointed the way toward the future of military training: interactive, immersive, complex, and variable scenarios in which the total experience of war could be brought forth in its digital replication. Because simulation was given much of the credit for the military's Gulf War success, the postwar period saw DARPA's *SIMNET*-related research and development efforts expand significantly.

The next major step in the military's video game history came with the 1993 release of the blockbuster first-person shooter fantasy *Doom.* According to Timothy Lenoir and Henry Lowood, historians of science, *Doom* is solely responsible for changing practically every facet of PC-based gaming, including "graphics and networking technology, . . . styles of play, notions of authorship, and public scrutiny of game content." (One of the game's innovations was a new mode of play called "death match," which, like *Doom*'s other innovations, is now a standard feature of many first-person shooter games.) *Doom* was an immediate sensation among gamers, with sales soon climbing into the millions.

Around the same time that *Doom* was released, the Marine Corps Modeling and Simulation Office (MCMSO) received a mandate from the annual General Officers Symposium to begin looking for commercial video games that might prove useful for training. Because the Marine Corps budget is a great deal smaller than that of the other services, the corps has a long history of seeking cost-effective training solutions. General Charles Krulak, its commandant at the time, believed that PC-based war games held great potential for teaching Marines critical decision-making skills.

Lieutenants Scott Barnett and Dan Snyder of MCMSO immediately began combing through dozens of military-related video games to see if any might be useful for training. They developed the online Personal Computer Based Wargames Catalog, on which they posted detailed reviews of the numerous games they investigated. Barnett and Snyder were looking for a fast-action first-person shooter — one that, crucially, allowed user modification. Of the many games they examined, only *Doom* (or rather its sequel, *Doom II*) fit the bill. As part of its marketing strategy, *Doom*'s developer, id Software, had released parts of the game as shareware and encouraged players to enact their own modifications.

Throughout the spring and summer of 1995, Snyder transformed the game from an outer-space gothic fantasy into a military fire-team simulation. The Martian terrain and alien demons of the original *Doom* were replaced by a dun-colored landscape of pockmarked concrete bunkers and enemies who had been drawn from scans of GI Joe action figures. The cost of production? A mere $49.95—the price of one copy of *Doom II*.

The point of the modified game, known as *Marine Doom*, was to teach Marines not how to fire their weapons but how to work together in teams and make split-second decisions in the midst of combat. "A real firefight is not a good time to explore new ideas," Snyder explains. The game had another, equally significant, rationale. "Kids who join the Marines today grew up with TV, videogames, and computers," Barnett reasons. "So we thought, how can we educate them, how can we engage them and make them want to learn?" Barnett and Snyder's calculations were correct: their creation became a huge hit among Marines, though, like *Army Battlezone*, it was never actually used for training. According to Barnett, Marines would plead to be allowed into his base's gaming lab even after it closed at night.

Marine Doom was created at a time when the Pentagon had begun embracing simulation for a broad range of activities. As scholar Sharon Ghamari-Tabrizi relates, these activities included "part-task training; mission rehearsal; operational planning; strategic and tactical analyses; weapons systems modeling during research and development, testing and evaluation, and acquisitions; and long-range future studies." Much of this emphasis on simulation was the result of post–Cold War military downsizing. With the collapse of the Soviet Union, the military's budget had been reduced to a level commensurate with what Congress assumed was a greatly reduced geopolitical threat. The relative affordability of simulation technologies matched well with the military's newly tightened budget.

The Federal Acquisition Streamlining Act of 1994 also forced the military to change its procurement policies. No longer could it underwrite defense contractors' R&D and acquisitions on an unlimited basis; instead, the Pentagon had to rely on what are known as "commercial off-the-shelf" (COTS) technologies—technologies that already exist and that have been developed by commercial industry. Take *SIMNET:* put together by military contractors, it required $140 million, ten years, and several hundred employees to build, even though it did use some commercial technologies. By contrast, *Marine Doom,* which relied exclusively on commercial technologies, was built by eight people in six months for $25,000. Military contractors now had to take on the stripped-down and flexible management practices of corporations—in effect, to become commercial businesses themselves. This had a deep and immediate impact on the defense sector, resulting in the merger or closure of a number of prominent companies.

In order to maintain their livelihoods, defense contractors had to find other customers to whom they could peddle their high-tech gadgets. Yet even in this time of seeming crisis, the contractors ended up coming out ahead, as it quickly became apparent that another industry was hungry for their wares: the entertainment industry. The relationship born of this outcome was symbiotic: defense contractors would spin their technologies off into the commercial game industry, and the commercial game industry would spin its technologies right back. In an update of Eisenhower's classic formulation, cyberpunk writer Bruce Sterling termed this win-win relationship the "military-entertainment complex"—the relentless exchange of technologies, personnel, and money that defines the bond between the military and the video game industry.

The military's reduced budgets in the 1990s also led to a greater dependence on reservist troops, which only increased the use of distributed interactive simulation systems for training. These systems enabled reservists to participate in large-scale training exercises and maneuvers no matter where they were based. As a further cost-cutting method, the immediate post–Cold War period saw the military's emphasis shift to all-encompassing joint operations, as opposed to individual service missions. Two new declarations of military doctrine—*Joint Vision 2010* (1996) and *Joint Vision 2020* (2000)—codified this focus. In the effort to develop new simulation platforms that would meet the requirement for a jointly linked system, the four services of the armed forces were directed to overcome their traditional rivalries. The ultimate result of this attempt at cooperation was the *Joint Simulation System* (*JSIMS*), a single, integrated virtual battlefield—in technical terms, a mission rehearsal and command simulation environment—in which participants from all four services could operate regardless of location.

There was one more reason for the military's turn to simulation: modern high-tech warfare was increasingly fought through electronic and digital interfaces resembling video games. Early on, this rapid growth in the electronic mediation of warfare caused confusion even among military professionals. An oft-repeated anecdote involves the war game *Operation Internal Look*, undertaken by the U.S. military in July 1990, during the run-up to the first Gulf War. General Norman Schwarzkopf relates the tale in his memoirs: "As [*Internal Look*] got under way, the movements of Iraq's real-world ground and air forces eerily paralleled the imaginary scenario of the game . . . As the war game began, the message center also passed along routine intelligence bulletins about the *real* Middle East. Those concerning Iraq were so similar to the game dispatches that the message center ended up having to stamp the fictional reports with a prominent disclaimer: 'Exercise Only.'"

Linking Entertainment and Defense

In the 1990s, no less an entertainment icon than Mickey Mouse presided over the tightening of the military–video game industry bond. At a mid-1990s meeting of the Army Science Board, that service's senior scientific advisory body, four-star general Paul Kern met Bran Ferren, an entertainment industry futurist with a friendly, expansive manner and a wild red beard. Ferren was the influential head of creative technology at Walt Disney Imagineering, the design and development arm of the Walt Disney Company based in Glendale, California. (Since its founding in 1952, Walt Disney Imagineering has developed dozens of innovations in the areas of special effects, interactive entertainment, fiber optics, robotics, and film techniques.)

General Kern's first thought upon meeting Ferren, with his tan explorer's jacket and his untamed facial hair, was, "What's this crazy liberal doing here in the middle of our organization?" As soon as he heard Ferren speak, however, Kern found him to be an inspiring, intellectually challenging figure who crystallized many of the nascent doubts Kern had been harboring about the static state of military simulation. Listening to Ferren describe the cutting-edge virtual-reality development efforts he was leading at Walt Disney Imagineering, Kern came to the abrupt realization that the entertainment industry had leaped far ahead of the military in regard to high technology — and, equally important, in the cost of that technology.

Ferren pointed out to the assembled army officials that when he and his entertainment industry associates looked at the military's modeling and simulation offerings, the offerings were, frankly, "unaffordable and pretty crappy." The software was "lousy" and the hardware was "clunky and inflexible." Ferren posed a series of questions to prod the army officials' thinking. "How much texture memory can we have in the graphics process unit?" he asked as an example. He received a number of blank stares. "Texture memory?" someone responded. "What's that?"

Kern's education was in mechanical engineering, and over the previous two decades he had gained a great deal of experience with computing and simulation, including the *SIMNET* program. He found Ferren such a font of valuable information and advice that he began meeting with him regularly, sometimes at Disney Imagineering headquarters, other times in his own office in the Pentagon. Their conversations were technical and wide-ranging, but Ferren took pains to drive home to Kern a simple message: "You gotta be where the action is." If the military wanted to be part of the emerging technology base in Hollywood, which was linked to Silicon Valley, then it had to establish a presence there.

Kern was so impressed by his meetings with Ferren that he charged his subordinates with making the military more Disney-like. The military had been at the forefront of technological development for decades, he told them. Why couldn't it now develop its capabilities to match those of the entertainment industry?

As it turns out, he wasn't the only one asking that question. In 1996, professor Michael Zyda of the Naval Postgraduate School in Monterey, California, had chaired a National Research Council study titled "Modeling and Simulation: Linking Entertainment and Defense." As tightly connected as the military and the game industry had always been, Zyda's report argued that the two entities nonetheless had a great deal more to offer each other. When General Kern sent his subordinates scrambling to find someone who could better meld the entertainment industry's technological know-how with the military's training needs, Zyda's name was at the top of the list.

An intriguing combination of laid-back Southern California surfer and overcaffeinated Silicon Valley entrepreneur, Zyda frequently uses words like "awesome" and "totally," yet those words are delivered with rapid-fire intensity and a buzzing physical energy. Compact and sturdy, with a gray mustache and thinning gray hair, Zyda is one of the people most responsible for the partnership between the military and the entertainment industry. While that link is hardly new—think of the numerous propaganda films produced by Hollywood during World War II—the mid-1990s witnessed the start of an unprecedented level of collaboration between the two groups. More than any other single person, Mike Zyda played a seminal role in this process.

Zyda's interest in computers dates back to his undergraduate years at the University of California at San Diego, where, as a freshman math major in 1973, he got a job working in the lab of a physical chemistry professor named Kent Wilson. The job interview was brief, with Wilson asking Zyda only whether he was willing to learn three things: computer graphics, programming, and how to write grant proposals.

Wilson's lab offered the kind of freewheeling intellectual and creative environment that marked the high-tech world at the time. Zyda worked alongside seventeen other undergraduates, experimenting with computers, lasers, and chemicals. Among his coworkers were Bud Tribble and Bill Atkinson, both now legendary in the annals of computing. Atkinson, the eleventh employee of Apple Computers, is the creator of MacPaint, QuickDraw, and HyperCard, while Tribble managed the original Macintosh software development team and helped design the Mac OS and user interface.

The "life-changing" experience of working for Wilson led Zyda to pursue a master's degree in computer information sciences at the University of Massachusetts at Amherst, where his adviser was Victor Lesser, a major figure in the field of simulation. After receiving his doctorate from Washington University in St. Louis, Zyda entered a job market that was remarkably ripe for new PhDs in computer science.

At the time of Zyda's graduation, many universities around the country were just beginning to create computer science programs, but the lack of graduates with relevant experience meant that there was a dearth of qualified faculty. Without even sending out applications, Zyda was recruited by the Naval Postgraduate School (NPS) in Monterey, California.

Zyda's focus when he arrived at the school, in February 1984, was real-time graphics. The army's *SIMNET* program had started the year before, and there was growing momentum in the military for visual simulators, which at the time cost between $10 million and $30 million each. In 1988 the army tasked Zyda with building a visual simulation system for the fiber-optic-guided missile. The FOG-M was an early version of a drone; it had a TV camera in the front and a thirty-kilometer fiber-optic cable spilling out the back. A soldier watching a video screen would guide the missile with a joystick and crash it into the intended target. Rather than relying on technology built by defense contractors, Zyda and his students built their simulation system to run on a $60,000 Silicon Graphics machine. They finished in six short weeks. When they presented their system to the army personnel at nearby Fort Hunter-Liggett, the response was immediate. "We're cutting you a check for $100,000," they told Zyda. "We want to take this system out into the field starting today."

Zyda and his students next built a simulator for the vehicle on which the FOG-M was mounted, after which they needed to network the two systems together. Luckily, Zyda had learned networking while on a three-week consulting trip in 1987 to Tokyo, where he had built a piece of code that would allow any number of workstations to be connected. He and his students now used that software to build a networked virtual environment that they called the "NPS Moving Platform Simulator." Soon after, Zyda received a phone call from a man named George Lukes at the U.S. Army Topographic Engineering Center.

"I just read a paper you wrote on your Moving Platform Simulator system," Lukes told Zyda. "It looks like *SIMNET.* Can I come to Monterey and talk with you?"

Zyda had never heard of *SIMNET,* because DARPA hadn't written any papers or given any talks on it. "What's *SIMNET*?" he asked, confused.

After sending Zyda a paper describing *SIMNET,* Lukes went out to NPS for a demonstration of Moving Platform. After the demonstration, Lukes took Zyda aside and told him how impressive it was. He then offered a proposal.

"Listen," Lukes said, "the army is just about to take ownership of *SIMNET* from the defense contractors, but no one in the military knows how to read and write the networking packets. There's also no one who knows how to read the terrain databases that the contractors have created. Will you do it?"

This was just the kind of challenge that Zyda and the students in his graphics class enjoyed. Using money provided by Lukes, they taught themselves how to read *SIMNET*'s packets and databases. The source code they created led them to build the Naval Postgraduate School Net, or NPSNET, a collection of Silicon Graphics workstations attached to a local-area Ethernet. NPSNET was in essence a *SIMNET*-connected simulator that enabled officers both to observe and to participate in the virtual training of their soldiers.

Zyda and his students' efforts quickly attracted the attention of numerous Department of Defense offices, all of which were interested in the training possibilities of virtual technology. In 1995, Zyda was asked to take part in a National Research Council (NRC) study called "Virtual Reality: Scientific and Technological Challenges," which advised the government on the kinds of virtual-reality research it should invest in. Though he was a relatively lowly member of the team, Zyda ended up writing about one-third of the final report.

Because of these efforts, the next year Zyda received another call from the National Research Council. The NRC had just received funding from Anita Jones, the Pentagon's director of defense research and engineering — who was responsible for overseeing the department's science and technology program, research laboratories, and DARPA — to put together a conference and a report on areas of potential collaboration between the defense and entertainment industries. Would Zyda be willing to chair the committee?

Anita Jones's interest in the subject stemmed from her previous tenure as chair of the University of Virginia Computer Science Department. There she had hired Randy Pausch, the computer science professor whose book *The Last Lecture,* published shortly before his death in 1998, became a national sensation. While teaching at UVA, Pausch had taken a sabbatical to go to Disney Imagineering in Orlando, where he worked on DisneyQuest, an indoor interactive theme park filled with virtual-reality attractions. Pleased with the results, Pausch invited Jones — who by then had moved to the Pentagon — to pay a visit to Orlando. As she toured DisneyQuest, Jones had a sudden realization: she was paying various Pentagon outfits heaps of money to build large-scale visual simulations, and yet what Disney had was far better and cheaper. This realization led her to fund Zyda's National Research Council conference and study.

The conference took place in Irvine, California, over two days in

October 1996. Two very different groups were involved. One consisted of military representatives from all four services, DARPA, the Defense Modeling and Simulation Office, and the Office of the Secretary of Defense. The other group consisted of entertainment industry personnel from such companies as Paramount, Disney, Pixar, and Industrial Light and Magic. Zyda, like Jones, wanted to capitalize on technological advances occurring not just in the military but in the worlds of entertainment and digital technology as well. While the specifics of these advances might have varied between fields, Zyda felt there was a key point at which they overlapped: simulation.

The conference featured testimony from the military side about its oft-failed attempts to do physically based modeling for virtual environments. The problem, the military people said, was that they would get wrapped up in the physics and the virtual environments would be difficult to upgrade. The entertainment group offered some simple advice. Look, they said, all you have to do is give people the *illusion* of an explosion happening; you don't have to do the actual physics. This was a wake-up call for the military folks, who, from the entertainment group's perspective, were trying to solve a bunch of problems that they didn't need to solve. Why not use games built by people who actually know how to build games, the entertainment people suggested, as opposed to using games built by defense contractors?

The Zyda committee's final report, "Modeling and Simulation: Linking Entertainment and Defense," claimed that by "sharing research results, coordinating research agendas, and working collaboratively when necessary, the entertainment industry and the DoD may be able to more efficiently and effectively build a technological base for modeling and simulation that will improve the nation's security and economic performance." In addition, the report declared it essential that academia be involved in this collaboration, arguing that the entertainment industry and the Department of Defense needed to band together to sponsor the development or further enhancement of academic programs dedicated to the fields of modeling, simulation, and virtual reality — all in the name of national security.

Not everyone was enamored of "Linking Entertainment and Defense." Anita Jones had delivered the funding for the study through the Defense Modeling and Simulation Office (DMSO). At the time, the DMSO was pushing something called the High Level Architecture, a new infrastructure for networking virtual environments and simulations across the entire Department of Defense. Testimony at the NRC conference pointed out several limitations of the High Level Architecture, but the DMSO insisted that Zyda's report praise it as the future for networking games. When Zyda refused, the DMSO was furious. After the study came out, Zyda asked Captain Jim Hollenback, the DMSO's director, what he thought of it. Hollenback did not mince words: "We hated your fucking report," he told Zyda. "We threw all the copies in the trash."

The End (and Beginning) of a Dream

Zyda had to wait two more years before the military was ready to accept his report's recommendations. In January 1999, he received a phone call from Mike Andrews, the chief scientist of the army, and Michael Macedonia, one of his former PhD students at the Naval Postgraduate School. Macedonia was now the chief technology officer at STRICOM, the Pentagon's simulation and training office. When General Kern had laid down his order to make the military more like Disney, Andrews and Macedonia were on the receiving end. They told Zyda they wanted him to write the operating and research plan for a new institute they planned to build at the University of Southern California, UCLA, or UC Berkeley. This facility, to be named the Institute for Creative Technologies, would give the army direct access to the game and virtual environment technology being developed by the entertainment industry and academia and would be funded by a Pentagon seed grant of $100 million. Andrews and Macedonia felt that "Linking Entertainment and Defense" provided the perfect road map.

The call couldn't have come at a better time for Zyda. After spending most of his career at the Naval Postgraduate School, he was looking for a way out. Working from "Linking Entertainment and Defense," Zyda wrote the research agenda and operating plan for the Institute for Creative Technologies in thirty days. He flew to USC to meet with the dean of cinema, the dean of engineering, and the director of the Information Sciences Institute. Then, in March 1999, Zyda went to the Pentagon to meet with Andrews and Macedonia in person. Both of them were enthusiastic about his document. "This is great!" they told him. "Why don't you go back and spend some more time socializing at USC? We want to build the institute there."

Zyda spent the next three months working on setting up the ICT. That June, however, Andrews and Macedonia abruptly stopped returning his e-mails and phone calls. He soon learned that the position of ICT director, which he had been promised, had instead gone to former Paramount television executive Richard Lindheim, a veteran of *Star Trek* and a close friend of USC dean Elizabeth Daly. Zyda had spent most of 1997 doing technology consulting for Lindheim, advising him on building the StoryDrive Engine for *Star Trek: Voyager.*

Denied the job he wanted, Zyda decided to create a research institute like ICT at the Naval Postgraduate School. Again using "Linking Entertainment and Defense" as a template, he set up the MOVES (Modeling, Virtual Environments, and Simulation) Institute, staffed by a combination of researchers and graduate students dedicated to modeling and simulation, with a core emphasis on computer gaming. This put Zyda in the unique position of building his own research institute to compete against the other research institute he had founded.

Both the ICT and MOVES ended up playing crucial roles in advancing the military's use of video games for training and education as well as for recruiting and mental health treatment. In later chapters we will see how the ICT and MOVES — along with the army's simulation and gaming office — are two key sites from which the twenty-first-century military-entertainment complex has expanded. Today the ICT in particular remains influential, and is helping to keep alive the military's tradition of technological innovation.

Yet the soil from which the military-entertainment complex has grown consists of more than just technology and video games. Equally relevant to this growth is the military's extensive, yet little noted, legacy of educational innovation. As we are about to see, this legacy—like that of technology—possesses surprisingly deep roots.

CHAPTER 2

Building the Classroom Arsenal: The Military's Influence on American Education

IN LATE DECEMBER 1777, General George Washington marched the tired, hungry, badly equipped troops of his Continental Army to Valley Forge, Pennsylvania, eighteen miles northwest of Philadelphia, to settle in for the winter. The choice of location was strategic: Valley Forge combined the elevation of Mount Joy and Mount Misery with the natural barrier of the Schuylkill River to form a secure location from which the army could keep watch on the British and ensure that the British could not launch any surprise attacks on their position. Yet the absence of attacking troops hardly meant that the army was safe from harm: starvation thinned its ranks, while proliferating diseases killed thousands more. Washington's repeated petitions to the Continental Congress for relief went nowhere; the Congress had no additional resources to give.

The harsh, wearing conditions of their winter camp threatened the Continental troops' morale and discipline — a potentially fatal situation for a ragtag army that was barely holding its own against the better-equipped and better-trained British units. The army was further hampered by its lack of a standard training manual. The American troops had received training, yes, but the variety of manuals used for this training meant that organized battle movements were nearly impossible to execute, a fact as potentially fatal to the troops as the lack of food and supplies.

On February 23, 1778, Baron Friedrich Wilhelm von Steuben, an elite former member of the king of Prussia's general staff, arrived at Valley Forge bearing a letter of introduction from Benjamin Franklin. Steuben wanted to offer his military skills to the rebel troops. Though the Prussian spoke little English, General Washington sensed his talents nonetheless and quickly named him acting inspector general in charge of creating and running an organized, unified training program. Steuben threw his considerable energies not only into drilling the men but also into composing a comprehensive drill manual, which his aides translated from French into English for each day's exercises. In addition to imposing uniform standards and methods on the Continental Army, Steuben broke new ground by working directly with his troops, breaking down the traditional barriers between commanders and their charges. By May of the same year, he had managed to turn a scruffy and disorganized army into a confident, even imposing fighting force.

Standardized battle drills weren't the only new educational event at Valley Forge that winter. Recognizing a need to provide basic literacy instruction to his soldiers, General Washington ordered that the Bible be used to teach them reading and writing. He wasn't looking to make his troops literate so they could perform their military tasks, but rather to enable them to read their Bibles so they might achieve greater spiritual awareness. His rationale was moral, not military, and he enlisted chaplains as instructors. Hardly noted in the years since, this seeming footnote to the storied history of Valley Forge was in fact the first instance of soldier education in the American military.

These two tracks first seen at Valley Forge — one involving training, one involving education — have had a major impact not only on the military but on the American school system in general. For the most part, the scope of this impact remains barely recognized. Yet some of the most dominant strands in American education, including standardized tests, adult education, and workplace learning, have been innovated, refined, or greatly expanded as a result of military needs and military funding. The military has also been a leading force in increasing access to education, especially for those from less privileged backgrounds. Because of this educational legacy, a full understanding of the military's current use of video games as learning tools is impossible unless one sees it, rightly, as the continuation of a long instructional tradition. During times of war, when large numbers of lower-skilled inductees have nearly overwhelmed the military's capacities, the armed forces have been a seedbed for new teaching methods and tools. In particular, military research and funding have been the primary drivers of educational technology since World War II, including the computers-in-the-schools movement that began in the early 1980s and that laid the groundwork for the welter of high technology in classrooms today. This is not to mention the myriad academic fields that owe their existence or their growth to military needs and funds: physics, both nuclear and otherwise—indeed, a great deal of science as a whole, including neuroscience, cognitive science, and information science; robotics; computer science and computer engineering; psychology, including behaviorism and human engineering; electronics, digital and otherwise; and earth sciences, including seismology, meteorology, and oceanography. As educational historian Douglas Noble points out, "The emphasis on science and mathematics education in the schools since Sputnik in the late 1950s is perhaps the most visible consequence of military technological enterprise."

Also influential have been the military's *concepts* of learning—often because civilian policymakers have failed to grasp the difference between training and education. Historically, for the military, learning is task-related, useful only as a means to achieving something else, something of strategic import. The armed forces are concerned not with knowledge for knowledge's sake but with knowledge for the performance of specific duties. This functional approach has spread from the military to our public schools, as a crucial part of what education historians W. Norton Grubb and Marvin Lazerson term the "vocationalism" movement, which they identify as "the single most important educational development of the twentieth century."

The military's influence on education can also be seen in regard to standardization, as the armed forces have remained at the forefront of standardized testing since the beginning. Even today the military continues to boast the largest testing program in the world. Here we see the origins of the idea that learning can be brought under scientific management, an idea that in the twenty-first century has been central to the high-stakes testing policies of the Bush and Obama administrations.

Since World War II, the military's technological advances have continually influenced national standards for literacy. Scholar Deborah Brandt writes that as military-funded technologies entered American society both during and after World War II, literacy came to be treated as an essential war-fighting resource and the crucial element in our postindustrial economy. Literacy became what Brandt calls "a collateral investment needed to get the most out of investments in technology."

The military's impact on American education isn't just the result of the fact that national defense plays such a prominent role in the life of any nation-state; it's equally due to the sheer size of the military's instructional needs. Every year, tens of thousands of new recruits enter basic training, after which they move on to over three hundred occupational specialties, along with numerous subspecialties, where they have to learn both battlefield and peacekeeping tactics. The vast majority of these new recruits are young and inexperienced, often fresh out of high school (or high school dropouts), and many of them have never before held a job. (The armed forces are the nation's largest employer of initially unskilled labor.) The U.S. military as a whole spends far more money on training — around 16 percent of the yearly defense budget — than any of the world's other militaries.

It's important to note, though, the manner in which the military's influence has often spread. In the aftermath of World War II, ex-military personnel often moved into the corporate or educational sectors, taking the lessons of their military experiences, insights, and preferences into their new jobs. Similarly, many of the main proponents of the computers-in-the-schools movement in the 1980s had military backgrounds or had worked on military-sponsored research projects and so were influenced by the military's interest in technology and computing.

The military's technological innovations spring from concrete problems that need immediate solutions. Take standardized testing: the armed forces, especially in times of war, have to classify and organize vast numbers of soldiers every year. (For example, over a two-year period in World War I, the army increased its number of personnel twentyfold.) These soldiers must be slotted into occupational areas for which they show aptitude. Whatever the inherent problems with standardized tests — and there are many — the military relies on them to accomplish such otherwise unmanageable tasks.

Other pressures include the military's relatively short training periods and its need for immediate results, coupled with the range and scope of its training mission. It can't just place a gun in someone's hand and say go; nor is firing a weapon the only skill a soldier needs, even in the infantry. Soldiers have to be trained and educated in basic skills (say, math), military skills (say, marksmanship), and specific job skills (say, operating complex weapons systems). They must be taught to perform these skills in less than ideal conditions (namely, war) and to operate as individuals, in teams, and in units. What's more, every soldier, and every unit as a whole, has to update and maintain these skills over years or decades. Efficiency, specificity, uniformity — these are the particular needs of military training, and they have long been reflected in the military's learning innovations.

The Twentieth Century: From Standardized Testing to Distance Learning

Born in 1876, Robert Yerkes grew up on a farm in rural Bucks County, Pennsylvania, and from an early age he knew that the hardscrabble life of a country farmer was not for him. After an uncle financed his way through Ursinus College, a small liberal arts college founded by members of the German Reformed Church, Yerkes accepted an offer from Harvard to undertake graduate study in biology. His stated goal was to become a physician, but he became sidetracked by an interest in animal behavior, which eventually compelled him to switch to the field of comparative psychology. After receiving a PhD in psychology in 1902, he took on a number of part-time jobs, including teaching at Harvard and working as the director of psychological research at the Boston Psychopathic Hospital. Fifteen years later, at the relatively young age of forty-one, when the United States had newly entered World War I, Yerkes became president of the American Psychological Association (APA).

At the time, the field of mental testing enjoyed little credibility, let alone standing, a circumstance that Yerkes was determined to rectify. Yerkes badly wanted to establish the field of psychology, barely a quarter-century old at the time, as one of the "hard" (that is, legitimate) sciences, and he believed that mental testing, with its reliance on quantification, would provide the rigor necessary to make psychology respectable in the eyes of the larger scientific community. He quickly perceived that the war would provide the perfect vehicle for popularizing these tests, and he urged the APA to become involved with the war effort. Nominated to chair the APA's Committee on Psychological Examination of Recruits, Yerkes approached army officials with the idea of administering intelligence tests to their incoming soldiers. It was not a tough sell: the army had already determined that new methods of mass screening were desperately needed to sort and classify the vast numbers of inductees.

To meet this need, Yerkes and a team of psychologists developed the Alpha and Beta tests, the armed services' first group examinations. The Alpha test, administered to literate English-speakers, was a multiple-choice written examination, while the Beta test, which did not use language, was intended for illiterates and non-English-speakers. The tests' formats would not look unfamiliar to test-takers today—they featured questions based on synonyms and antonyms, analogies, and the unscrambling of sentences. The results of the Alpha led to cries of a national literacy "problem": of the 1.7 million men who took the exam, 30 percent could not read the form well enough to understand it. Given that the majority of these men had had some type of formal schooling, the results alerted educators nationwide to problems with reading instruction in the schools, the first time that methods of teaching literacy had been debated on such a large scale.

From the army's perspective, the tests' benefit was that they provided supposedly objective criteria with which soldiers could be classified, trained, and placed into appropriate jobs—including, importantly, officer training. The tests could also be used to identify soldiers with low skills who, provided they received remedial training, might still perform some useful function. And while intelligence testing convinced some cynics that many adults weren't mentally capable of benefiting from education, the war years also proved that thousands of people thought uneducable could gain basic reading skills in a mere six to twelve weeks. By 1919, almost 25,000 illiterate and nonnative personnel had received this quick-fix instruction.

The large amount of data collected by Yerkes and his colleagues was ultimately used to more pernicious effect than their military mandate called for. Manipulated and inaccurately analyzed in a way that supported racist and eugenicist beliefs for decades to come, the data, as Stephen Jay Gould wrote in his classic work *The Mismeasure of Man,* gave rise to three "facts" that greatly influenced American social policy:

1. The average mental age of white American adults [stood] just above the edge of moronity at a shocking and meager thirteen;
2. European immigrants [could] be graded by their country of origin . . . The darker peoples of southern Europe and the Slavs of eastern Europe [were] less intelligent than the fair peoples of western and northern Europe;
3. The Negro [was] at the bottom of the scale with an average mental age of 10.41.

To the major proponents of testing at the time, almost all of whom were ardent eugenicists, the data seemed to verify their extant racial and cultural prejudices—prejudices that continued to be built into the structure of standardized testing throughout the twentieth century and on into today. Thomas Sticht notes that the army Alpha and Beta tests also "produced a way of thinking about intelligence and aptitude that has continued to underpin the use of mental tests" in the military and in our public schools, including the basic concept of *mental categories* and the idea that standardized tests can reliably sort people into the various categories.

World War I also marked the first time that specialized training was a major military priority, as it became clear that soldiers needed to operate increasingly complex equipment. During the Civil War period, over 90 percent of enlistees had been engaged in nontechnical combat-related activities, while less than 10 percent of the force worked as craftsmen, clerical personnel, or technical personnel. By World War I, however, fewer than 50 percent of enlistees performed nontechnical combat-related duties, while skilled and semiskilled personnel were needed in great numbers. For this reason, adult and vocational training became a major concern for the armed forces during World War I, when technical specialists were in short supply. (Fewer than 20 percent of the necessary specialists were available without training.) This prompted the army to adopt a functional approach to literacy, in which lessons focused exclusively on job-related tasks. Comprehension, not just decoding, became the goal, as soldiers now had to be literate enough to read text-based manuals relating to their jobs. This resulted in a number of experimental programs, such as one at Camp Grant in Illinois, where soldiers received lessons in reading, math, and civics along with technical and vocational training. The declared aim of these programs was "to develop arrested mentality" as quickly as possible.

The military's adult education efforts were bolstered by the National Defense Act of 1916, which ensured that soldiers would "be given the opportunity to study and receive instruction upon educational lines of such a character as to increase their military efficiency and enable them to return to civil life better equipped for industrial, commercial, and general business occupations." The war effort also led to the passage of the Smith-Hughes Act in 1917, which greatly expanded vocational training in high schools nationwide.

After World War I, the attention generated by the army's wartime experience, along with the growing eagerness of school administrators to embrace science-based methods of organization, led to the growth of standardized achievement testing as a mass educational movement. The first Scholastic Aptitude Test (SAT), given in 1926, was a modified version of the army's Alpha exam — an unsurprising development, given that the well-known eugenicist Carl Brigham, a key figure in developing the army tests during World War I, was tapped by the College Board to develop the SAT. Many of the original test's questions made the military connection explicit, as in this math problem: "A certain division contains 5,000 artillery, 15,000 infantry, and 1,000 cavalry. If each branch is expanded proportionately until there are in all 23,100 men, how many will be added to the artillery?"

While the bulk of the army's instructional efforts ceased after the end of World War I because of the drawing down of American forces, scattered efforts to train illiterate soldiers remained. Using army booklets as source material, the lessons emphasized reading and writing, although the booklets' content concerned history, civics, basic hygiene, and other topics aimed at enabling soldiers to become "productive" members of society. In this same period, the use of scientific personnel research for selection and classification mostly fell by the wayside, as the military contracted to its prewar size. At the beginning of World War II, the army inducted soldiers based merely on the inductee's own assertion that he could understand simple orders given in English.

Following the attack on Pearl Harbor, tests were brought back into play in order to screen the vast numbers of new soldiers, not only to determine their fitness for military service but also to slot them into aptitude-appropriate jobs. The Army General Classification Test (AGCT) took the place of the old Alpha. (Although intended as an examination of general learning ability, norms for the AGCT were based solely on the responses of white male soldiers and Civilian Conservation Corps members.) As the war continued, the large numbers of illiterate and semiliterate inductees spurred the development of additional unwritten exams. Between 1941 and 1945, the years of U.S. involvement in World War II, almost every important American in the field of testing was involved with the military in some way.

World War II also fueled the expansion of standardized testing by generating a need for the kinds of skilled and educated people who, once recruited, would be key military assets. This in turn focused attention on the importance of college study. With the enactment of the G.I. Bill in 1944, tens of thousands of veterans entered college for the first time. The sudden appearance of so many new students greatly increased the appeal of the SAT for school administrators, owing to the efficiency of its multiple-choice format. Though unintended by its creators, the G.I. Bill in effect did away with the elitist notion that blue-collar personnel were not fit for college study and that college should be reserved for the privileged.

World War II brought other instructional changes as well. The large numbers of new personnel in the war, for example, required a training model based in large part on group learning in classroom environments. To facilitate this model, military psychologists made significant advances in the development and use of educational technology. These psychologists also examined the military's existing learning principles to determine whether they really worked in the training realm. From this examination came a newfound military focus on such principles as "part-task training," in which complex tasks are analyzed and then broken down into components that can be mastered more easily. This process is today a crucial part of the civilian "vocationalism" movement, which, as its name implies, emphasizes vocational training in public education.

Between 1941 and 1945, the military launched a program of adult basic education whose scale was perhaps unrivaled in human history. This program established permanently for the armed forces the idea that adult learning could significantly improve job performance. It also made education relevant to soldiers by giving them academic credit for knowledge gained in the workplace, another innovation that quickly made its way into the civilian world.

Given the relatively short time period (twelve weeks at most) in which soldiers received instruction during the war, the focus was on specific military-related content knowledge, and instructional materials were tailored to a fourth-grade reading level. At the same time, the criterion by which an "acceptable" level of literacy among soldiers was defined fluctuated widely depending on manpower needs. Between 1941 and 1945, the minimum standards for enlistment underwent constant revision; the more recruits the army needed, the more it depended on evidence of educability, rather than level of education, in determining who could join its ranks.

The increased military emphasis on education during World War II also resulted in the development of General Educational Development (GED) tests, which enabled soldiers to use their military experience to qualify for high school equivalency degrees. Initially designed by the staff of the United States Armed Forces Institute (USAFI), the tests were at first restricted to military personnel and veterans. Beginning in 1947, however, the GED was offered to civilian adults; by the end of the next decade, more nonveteran than veteran adults were taking the test. Today, of course, the GED is used widely throughout North

America for the purpose of high school equivalency certification. In the United States, it accounts for almost 15 percent of annual high school diplomas.

The United States Armed Forces Institute was important for another reason, too. Tasked by the War Department to be a correspondence school for enlisted men, and using the U.S. Postal Service at home and the army post office system abroad, the USAFI pioneered the large-scale use of distance education. Study by correspondence was viewed as an efficient and attractive alternative for soldiers stationed overseas: there was no set starting or ending date, students could work at their own pace, and the courses were equally appropriate for individuals and groups. By the end of World War II, the USAFI was operating branches in places as varied and far-flung as Puerto Rico, Anchorage, London, Rome, Manila, New Delhi, Cairo, and Panama.

The Development of Computer-Based Learning

During World War II, the military's focus on literacy stemmed less from the war's expanding reach than from crucial and complex changes in the actual conduct of war. As Deborah Brandt relates, military-related advances required soldiers to "mediate technologies, gather intelligence, operate communication systems, and run bureaucracies that were growing faster and more elaborate" at every turn. Yet the technological advances of World War II didn't just drive the military's literacy standards upward — they also, and more profoundly, "changed the rationale" for education, from one of morality to one of productivity. Literacy had long been conceived of as a tool for social and religious stability: it enabled citizens to read their Bibles and served to acculturate and "tame" the immigrant masses. But during the war, literacy was detached from its moral associations. It became instead what Brandt calls "a needed raw material in the production of war." As a result, education was transformed from "an attribute of a 'good' individual" into a resource "vital to national security and global competition."

The technological basis of this shift has played out in the military's post–World War II learning innovations, as the military, more than any other single institution, has forged the link between education and technology. For decades the armed forces have been the world's most significant funders and developers of computer-based instruction and educational technology, both independently and through partnerships with industry. As one account puts it, "Computers would probably have found their way into classrooms sooner or later, but without [ongoing military support] it is unlikely that the electronic revolution in education would have progressed as far and as fast as it has." In addition to computers, the military has funded multimedia applications, simulations, instructional films, instructional television, overhead projectors, intelligent tutoring systems, teaching machines, and language laboratories. Not all of these were invented by the military, but all were developed, refined, and popularized by the military.

Computer-based education has its roots in World War II–era military research into man-machine systems — that is, integrated systems of humans and machines. Historian Martin van Creveld outlines the questions that in the 1940s and '50s captured the imaginations of military engineers and psychologists: "Which are the strong points of man, and which are those of the new machines? How . . . should the burden of work be divided among them? How should communication between man and machine . . . be organized?" These inquiries lay at the heart of the military's initial forays into computer-based education.

In 1958, behavioral psychologist B. F. Skinner published an influential article, "Teaching Machines," that attracted wide notice in the military and industry alike. (A teaching machine is a device used for automated instruction.) That same year, the American Psychological Association and the Air Force Office of Scientific Research held symposiums on the topic within four months of each other. Not coincidentally, the Defense Advanced Research Projects Agency was founded in 1958. Conceived of as a response to the perceived Soviet threat, DARPA provided essential funding for research in computer-based instruction.

The most notable project of the era was PLATO (Programmed Logic for Automated Teaching Operations), based at the University of Illinois at Urbana-Champaign. Funded by the air force, army, and navy, PLATO was a computer system designed expressly for educational purposes; researchers wanted to highlight both the pedagogic and the economic advantages of computer-based instruction. In a pioneering turn, PLATO had a plasma screen that offered text and rudimentary animation. The project also led to the development of a programming language for educational software. For years PLATO was the world's most widely used computer-based instructional system.

Many of the other most significant advances in computer-based education during the 1950s and '60s derived from the air force's Semi-Automatic Ground Environment, or SAGE. From this project came significant advances in core memory, keyboard input, graphic displays, and digital communication over telephone lines. The SAGE system also pioneered the design of "user-friendly interfaces" — think, for example, of help menus or online instruction aids — that teach users how to work with a particular system. On a more abstract level, the SAGE program fostered the idea that computer-based systems could be used as tools to enhance cognition. In this model, the functioning of the human brain was reconceived in the image of computer processing, which ultimately led to the field of cognitive psychology. Decision-making and problem-solving, in both real-world and simulated environments, were the focus of SAGE's computer-based instructional efforts. Altogether, SAGE marked the world's first example of computer-managed instruction.

The Education Gospel and Technological Change

As this chapter shows, the military has had a significant influence on American education. Channeled through a variety of intermediaries, especially the corporate world, military-sponsored methods, concepts, and technologies have repeatedly ended up in our public schools. The military's longstanding technological focus also explains why it has been a driving force behind what W. Norton Grubb and Marvin Lazerson term the "education gospel," the social assumptions surrounding the postindustrial transformation of public education's purpose. This is how the authors describe the gospel's vision: "The Knowledge Revolution (or the information society, or the high-tech revolution) is changing the nature of work, shifting away from occupations rooted in industrial production to occupations associated with knowledge and information. This transformation has both increased the skills required for new occupations and updated the three R's, enhancing the importance of 'higher-order' skills, including communications skills, problem solving, and reasoning." Without the military's seminal role, this transformation would not have assumed the same shape and form nor happened so quickly.

In keeping with this transformation, the authors note, workers today are required to engage in the kind of lifelong learning that enables them to keep up with rapid technological advances. The military has shaped this larger societal shift by providing a great deal of the technical apparatus and institutional rationale behind it. Decades ago, sociologist Daniel Bell noted that military technology was the "major determinant" of what he later termed the "information society." Taking his cue from Bell's analysis, Douglas Noble argues that the military's influence on "information theory, systems analysis, nuclear energy and transistors, . . . automation, robotization, [and] bioengineering" made it the "advance guard" of our high-tech economy.

The most significant effect of the military's learning-related efforts in the twentieth century was tying education and advanced technology inextricably together — not just technology in the sense of electronics and fiber optics and digital files, but technology as the center of our economy. This is not to suggest some Pentagon-driven conspiracy; any number of issues and players have influenced the link between technology and education. But the military's seminal role in this link is both essential and ongoing.

Remaining on the cutting edge has not been without its pitfalls. In practice, while the military's efforts at instructional innovation have led to a number of lasting changes, there have been numerous cases in which the promise and hype of new educational technology has far exceeded its actual benefits. Yet such recurrent disappointments have not dampened the military's enthusiasm for technology-driven instruction.

There is, however, a crucial difference today: for the first time in its history, the military wants to teach even junior personnel not just *what* to think but *how* to think. In the past, as we've seen, innovations have primarily been in the service of functionality and standardization. That's what makes current developments so remarkable: they break this mold. The complexity of the wars in Afghanistan and Iraq forced the military to accept that functional training for specific military purposes was no longer enough. As we will see, this is one major reason that the military is now investing so heavily in video games.

CHAPTER 3

"Everybody Must Think": The Military's Post-9/11 Turn to Video Games

And for pleasure, there was the simulator, the most perfect video game that he had ever played. Teachers and students trained him, step by step, in its use . . . It was exhilarating to have such control over the battle, to be able to see every point of it.

— ORSON SCOTT CARD, *Ender's Game,* 1985

IN THE LATE 1990S, with the Cold War still fresh in its coffin, military officials began to hypothesize that the skills and attitudes of modern teenagers differed from those of their elders in ways that would benefit the fighting of wars against America's new stateless enemies. A 2001 Army Science Board study laid out the specifics of these skills and attitudes. Teenagers, the study said, were excellent multitaskers; they could listen to music, talk on a cell phone, and use a computer at the same time. They preferred concrete reasoning, not abstract, deductive philosophizing. No longer did teens want to learn through passive listening; they wanted their education to be hands-on, experiential, based on specific examples. Even their definition of literacy had changed; it no longer referred to texts alone but also referred to images and multimedia. All these attributes, officials claimed, were relevant to the technology of modern warfare — and were fostered by playing video games.

Not coincidentally, these were the same skills and preferences considered essential to Defense Secretary Donald Rumsfeld's "revolution in military affairs" — otherwise known as "transformation" — which held that the U.S. military's high-technology combat systems and heavy reliance on air forces had dramatically reduced the need for large numbers of troops on the ground. In 2001, as the George W. Bush administration swept into Washington, "transformation" became the Pentagon's buzzword, the presumptive answer to post–Cold War uncertainties. The future of war was not lumbering tank battles between two superpowers; it was "asymmetric warfare," battles between opponents of vastly different strengths and capabilities. Transformation required a wholesale technological upgrade of the armed forces, with the goal of changing the military into a lithe, agile, easily portable fighting force that could be instantly deployed to any of the world's future hot spots. To do so would require "jointness" and "networking" — in other words, extreme cooperation between the military's four services, all of which would be connected by advanced technology. The events of 9/11 only increased Rumsfeld's push for a wholesale reduction of ground forces, one of the key elements that would lead to such disastrous results in Iraq.

And yet "transformation" did get one thing right: in the twenty-first century, the role of American soldiers had indeed become more complex and was being driven by advanced technology at every level. As a result, the military began to emphasize the battlefield importance of "situation awareness": recognizing that important information was available, understanding and interpreting the information's relevance to the mission, and using that information to forecast future plans and events. According to Jim Korris, former creative director of the Institute for Creative Technologies, the military "decided that it needed to think less about educating people on the physics of artillery tubes and start teaching them how to make smart discriminations very quickly in close urban fights — training in cognitive decision-making." Adaptive thinking, collaboration with others, system and information management: these competencies became (and remain) the order of the day. The belief that soldiers needed only the skills required to comprehend field and weapons manuals was superseded by the drive for digital expertise, for the highly advanced information-processing capabilities that video games supposedly promote.

This belief was highlighted in the Pentagon's 2003 study "Training for Future Conflicts," which argued that American soldiers needed to develop their thinking skills as much as their weapons-handling skills. Dr. Ralph Chatham, coauthor of the study and at the time a program director at DARPA, wrote that twenty-first-century warfare had increased "the cognitive demands on even the most junior levels of the military." As the report more bluntly stated elsewhere, "Everybody must think." At the time this represented a significant break from many of the calcified traditions of military training. In the principles it outlined and in the scope of its influence, "Training for Future Conflicts" laid the groundwork for the military's new game-based approach to learning.

The Rise of DARWARS

In his report, Ralph Chatham emphasized the idea that military proficiency depends as much on the soldiers who are operating today's complex technological systems as it does on the technology itself. (This was a continuation of the military's long-established man-machine emphasis.) Chatham said that the "speed, degree, and duration of complicated cognitive tasks" that soldiers were expected to perform meant significant improvements in military training were required. What might this mean in practice? The answer, Chatham claimed, lay in popular culture. Specifically, he argued that contemporary military training should be modeled on the massively multiplayer online video game (MMOG) world, where hundreds of thousands of participants play simultaneously in virtual environments.

The relevance of this model can be seen in Dutch scholar David Nieborg's analysis of the game *America's Army*. Players, Nieborg points out, are required to manage the game's multiple streams of data, keeping in mind such questions as "where are the opposing forces, who is talking to me, where are my teammates, where is my medic, how much time do I have left to complete the mission, how many bullets do I have left, what is the quickest way out of this hospital, what is that noise[?]" These questions and more dominate every moment of game play, indicating just how complex MMOGs can be.

For Chatham, writing in the wake of the American invasion of Iraq, MMOGs represented the next leap forward from the simulation-based combat training centers the armed forces had used in the 1970s and '80s. This time around, the focus would be on teaching soldiers how to process and react to information. Chatham's vision of overhauling military training through video games was the product of his uniquely extensive experience in that realm: during his time at DARPA, he had led the development of the groundbreaking DARWARS program, a set of digital "universal, on-demand, persistent training wars" that was used from 2003 to 2008 on U.S. military bases throughout the world. Though the graphics were a step below those of a commercial video game, they were far better than those in any previous military venture. Among the areas emphasized in DARWARS were cross-cultural communication, convoy operations, infantry tactics, and rules of engagement. The program focused on individualized instruction, direct feedback on performance, just-in-time training, and collaborative and self-paced learning—the same elements highlighted in Chatham's Army Science Board study. In a pioneering twist, which represented a significant development from earlier military simulations, soldiers could create new scenarios in DARWARS based on their own battlefield experiences. DARWARS showed a generation of military officials that video games could be uniquely effective training tools for America's twenty-first-century wars.

One of the most popular DARWARS tools, for example, was the game *Tactical Iraqi,* which trained soldiers in Baghdad Arabic and Iraqi culture. Chatham says the game — and the larger Tactical Language and Culture Training Systems program, which is still used today by both the army and the Marine Corps — stemmed from the belief that in today's wars, "nonverbal messages are as important as verbal ones." American soldiers in foreign countries are supposed to become as proficient in basic cultural and gestural cues as they are in basic vocabulary. In *Tactical Iraqi,* players interacted with virtual Iraqi civilians in various states of distress, with the success of the interaction depending on how players navigated both major and minor details of cultural communication. (For example, a thumbs-up gesture was highly offensive to Iraqis, as was a soldier's failure to remove his or her sunglasses before addressing civilians.) The game's emphasis on training even junior personnel in cultural difference represented a new phase for the military; in the past, understanding the culture in which one was fighting would have been seen as the purview of the State Department, not the armed forces.

In Chatham's view, decoding commands on today's battlefields had become only the most basic prerequisite for competence. Video games, he argues, are perfect for teaching soldiers what has become the essential twenty-first-century battlefield skill: how to "receive, triage, assess, decide, and act on masses of incoming data," even as "the sight of your buddy" is replaced by "icons on a screen."

From DARWARS to Real World

This refrain, which emphasizes the data overload that typifies contemporary warfare, explains why the post-9/11 military first grew so interested in gaming's cognitive benefits. Early on, Colonel Casey Wardynski, creator of *America's Army,* described the new challenges for American soldiers: "Our military information tends to arrive in a flood . . . and it'll arrive in a flood under stressful conditions, and there'll be a hell of a lot of noise . . . Most of it hasn't been turned into information and it's still data . . . How do you filter that? What are your tools? What is your facility in doing that? How much load can you bear?" At the time these questions were bedeviling military thinkers and planners struggling to adapt to the realities of war in Afghanistan and Iraq.

Wardynski, for one, argues that the gaming generation possesses an innate ability to dominate visually chaotic space. This is a key advantage against today's enemies, he says, who are not "stupid enough to chase us out into the desert where we can get them" but instead are going to hide in cities, where all is "chaos and clutter, and there's noncombatants and all that junk." Against this chaos and clutter, somebody who can pick a key piece of information out of a scene and decide in a split second what to do about it is a strategically important person for the army.

What DARWARS and *America's Army* and similar games gave soldiers, Wardynski says, was "a *virtual* classroom . . . a place in the world where you can take your mind—your body can't go because of distance, time, money, or danger—and you can separate your experiences from the limitations of the world." His vision proffered a high-speed human-computer interaction as the only viable tool for delivering effective instruction. The point of this instruction, like its method of delivery, was different from that in previous wars: its goal was to significantly improve the cognitive capabilities of recruits. "It's not a traditional way of thinking about training," Wardynski notes. "We usually think about training in terms of, 'I'm going to give you a set of answers, and a set of circumstances; you're going to memorize them, you're going to regurgitate them in the right order and in the right circumstances.' The [video game] way of thinking about training is, 'We're going to train your brain to better process what you take in, no matter what it is.'"

This notion of training the brain was a response to what the military termed the "three-block war," the paradigm of battle outlined by General Charles Krulak after the United States' experience in Bosnia,

Somalia, and Kosovo. As he explained (in the video game *Full Spectrum Warrior*): "In one moment in time, our service members will be feeding and clothing displaced refugees — providing humanitarian assistance. In the next moment, they will be holding two warring tribes apart — conducting peacekeeping operations. Finally, they will be fighting a highly lethal mid-intensity battle all on the same day. All within three city blocks."

Dan Kaufman, project manager of the DARPA video game *RealWorld* — the follow-up to DARWARS — made this same point: in the twenty-first century, American soldiers were no longer just war fighters. Adding to the pressure, Kaufman said, is that while following the chain of command used to be effective, there was no longer time for it. Soldiers in Afghanistan and Iraq had to be able to act and react immediately, and so they needed to make decisions that in the past would have been made by commanders. Kaufman, like Chatham and Wardynski, was one of the officials trying to figure out how to prepare soldiers for this situation — and coming to the conclusion that video games were the answer.

Kaufman may have been biased in this conclusion: he was a long-time veteran of the commercial game industry and had joined DARPA to apply the knowledge he'd gained there to a military context. Kaufman argues that games like *RealWorld* can make the cognitive load on soldiers more instinctive, like a kind of sixth sense. "Not in a mystical way," he says. "A sixth sense is just experience and awareness." He points out that there are different kinds of intelligence. Although a professor of physics, for example, might have the edge in a physics lab, "who would you want next to you if you were suddenly thrown into the middle of Baghdad — the physics professor or the nineteen-year-old kid who has been playing video games all his life?"

Kaufman acknowledges that the nature of the all-volunteer force means that some soldiers possess substandard reading and writing skills. Lots of people join the military, he says, because they did poorly in school, or because they hated school, or because they simply couldn't get other jobs. Video games, he argues, are often better matches for the particular skill sets and preferences of a large swath of today's recruits.

Still, Kaufman notes that soldiers in training are honing themselves for very specific, very difficult jobs, ones in which they need skills that earlier methods of instruction can't deliver effectively. Soldiers want to come home safely, he says, and video games are faster and better at helping them learn those skills. It comes down to what is "compelling," he reasons. Soldiers are compelled to use the most efficient tools "for teaching them how not to get killed."

What You Can Do with a Six-Year-Old

Along with Chatham, Wardynski, and Kaufman, one of the most prominent architects of the military's post-9/11 game use was Michael Macedonia, whom we met in Chapter 1. In the early years of the twenty-first century, Macedonia was the chief scientist and technical director of the army's gaming and simulation office, PEO STRI (the Program Executive Office of Simulation, Training, and Instrumentation). During his stint at PEO STRI (then called STRICOM), Macedonia's job was to guide army investment in simulation- and video game–based training technology, which fit well with his love of science fiction and computers. He used his bully pulpit to publish influential papers with titles like "Games, Simulation, and the Military Education Dilemma." A student of military history, Macedonia argued at the time that the military's use of video games was but the latest manifestation of a centuries-old tendency: "People have been using simulations for thousands of years, as long as there's been a military. They told stories, drew pictures in the sand, invented chess . . . They made these abstractions in the hopes that they could understand the nature and dynamics of war. If you look at what a scientist does with mathematical equations, what an artist does, or a writer — they're trying to abstract the universe."

Today Macedonia is vice president of the defense contracting giant Science Applications International Corporation. From this powerful perch he remains a strong advocate of the military's use of video games, in part because he thinks games overcome the key obstacle in soldier preparation: a lack of time. Time, he says, is "the tension in education. At one point in the history of mankind you could send somebody to school for two years and know everything there was to know. Nowadays, just college itself is four years. So with military training, time is the big problem. If I had enough time, I could prepare you for anything." This issue is hardly unique to the military, but the life-and-death stakes in that profession raise the ante on trainers and soldiers alike. "Colleges don't care whether you succeed in four years or eight years. The military can't afford that," Macedonia says. "I've been in arguments where they have argued about one day in basic training for a specific kind of training that everyone said was needed. There's always an argument about how much time do we need to get soldiers physically fit, how much time do we need for chemical-biological training, how much time do we need to make sure the soldiers can run UAVs [unmanned aerial vehicles]." Video games enable the military to compress the learning process.

In Macedonia's telling, the immersive nature of video games not only adds to their believability but also increases their storytelling power, one of the elements that gives them their training edge. Emotion, he says, "is critical to learning, and one of the key aspects of eliciting emotion is being able to provide a story. It goes all the way back to Homer. Look at *The Iliad* and the oral tradition — that was the way history was taught. The only way to remember all these facts was you put them into a story. A story is a way for folks to be able to understand, to absorb, and to retain."

Part of the military's (and Macedonia's) original inspiration for using video games in fact stemmed from a story: Orson Scott Card's young adult novel *Ender's Game.* In the novel, the six-year-old protagonist, Andrew "Ender" Wiggins, is a student at the national military academy, where he spends his days playing what he thinks is a virtual-reality game in which he is a soldier defending the earth against alien attack. We ultimately learn, however, that the battles Ender fought weren't simulations at all — they were instead quite real, and Ender's skillful fighting has just saved the earth from alien invasion. Macedonia says that the book was a major influence on the military's thinking about video games.

Macedonia is also a devotee of UC San Diego neurologist V. S. Ramachandran's theory that human beings consist, in essence, of their memories. What military training tries to do, Macedonia explains, is *create* these memories in soldiers before they ever hit the battlefield. He cites Ramachandran's notion that human beings have a virtual-reality program in their minds. "Somebody throws a ball at you; you anticipate where that ball is going to be. Really, what you are doing is running a little simulation of the world through your head." That is what the military tries to do with video games and simulations, he argues.

If all this sounds a bit theoretical, Macedonia insists on the practical nature of his vision. Ask him whether the military is expecting too much from video games, and his response is an emphatic no. The real question, he says, is, "What would you do otherwise? That comes up in our company all the time: 'Do you have an alternative?' Usually the answer is no. Unless the military is somehow able to recruit PhDs who come out of universities and are also really buff, and who speak three languages, particularly Urdu, and have substantial experience in foreign countries. No — you are dealing with eighteen-year-olds, and you have limited amounts of time."

Macedonia's ideal solution would be a fully encompassing virtual environment in which soldiers were always training. In this environment, he could have "soldiers always be[ing] part of the game . . . real people in real places interacting with real people in virtual places that are copies of the real world." He acknowledges the bizarre nature of this concept. "It does get really weird, and really kind of becomes science fiction at a certain point. It really is *Ender's Game.*" And yet to Macedonia that is just as things should be. After all, he says, "I've always been fascinated by what you could do with a six-year-old."

Literacy, Warfare, and Society

Science-fiction fantasies aside, defense officials emphasize that the military's use of video games is linked in a broader sense to changing notions of literacy within the services and in society as a whole. While reading and writing have long played an important role in warfare — for record-keeping, information distribution, propaganda, and maintaining archives — their value only increased during the bloody world wars of the twentieth century, as sophisticated new technologies placed greater demands on soldiers' literacy and the bureaucracies responsible for managing armies grew exponentially larger and more complex. In the United States, the military came to depend on literacy as a critical resource for rating, sorting, classifying, and placing its soldiers. Moreover — and this is a trend that has accelerated in America's recent wars — information itself became increasingly important as both a tool and a resource for war.

We saw earlier that over the past century the military has exerted a powerful influence on literacy, largely by sponsoring the development of cutting-edge technologies. People then had to gain the literacy skills needed to run these technologies, which also transformed economic and social systems.

As new technologies and duties have emerged, the definition of literacy has changed to encompass whatever skills are needed to handle them. And as the military has been quick to recognize, video games represent one of the more culturally prominent examples of a new sort of literacy. Although the press and politicians have routinely characterized video games as harmful, or simply ignored them, a growing number of education scholars argue that the various moves entailed in playing video games match up quite well with those involved in more traditional forms of literacy. They also match the skills needed to succeed in a complex, high-tech workforce, these scholars say.

Though the notion that video gaming represents an influential new literacy and is part and parcel of the broader ascendance of digital literacy skills was controversial when it was first introduced in the early years of this century, there is now broad consensus among educational theorists that it is valid. Foremost among those connecting literacy and video games is the distinguished literacy scholar James Paul Gee, who argues that contemporary video games are not only lengthy and intricate, they require players to learn and understand complex systems of words, symbols, problems, and cues. Gee claims that we must start thinking of literacy in this broad way, instead of only as the ability to read and write, because written language today is but one of several important modes of communication; images, symbols, sounds, and movement can be equally significant. Video games, he says, are a prime medium where these elements join together.

Constance Steinkuehler, a former senior analyst in the White House's Office of Science and Technology Policy and a professor of education of the University of Wisconsin–Madison, shares Gee's view and has drawn even more explicit links between gamers' skills and established educational standards. Video games require players to understand and engage in a densely "literate space" of icons, symbols, gestures, actions, visuals, and text. As Steinkuehler notes, "If we compare what individuals do within these spaces to national reading, writing, and technology standards, it turns out that much of their activity can be seen as satisfying these standards. For example, as recommended by the National Council of Teachers of English standards, gamers 'read a wide range of print and non-print texts' to build an understanding of texts and of themselves; use a wide range of strategies to 'comprehend, interpret, evaluate, and appreciate texts' . . . ; 'gather, evaluate, and synthesize data from a variety of sources' . . . ; and 'use . . . visual language to accomplish their own purposes.'"

Of course, even in this digital age, not all teenagers are computer wizards. According to a U.S. Army Research Institute study, there exists "a diverse population of soldiers, one that has individuals with limited computer skills to individuals with programming skills." This diversity in skills means that all digital training must be adaptable enough to train both highly skilled people and those with low skills. This reality isn't always reflected in the media, however. By far the most common sentiment is expressed by *Wired* magazine's Steve Silberman, who writes that today's soldiers have in many ways been training for their missions "all their lives. They pound on Halo in the garrison and launch strikes on Game Boys while riding in tanks. On their days off, they pile into the multiplex to see blockbusters crafted by the same technicians of verisimilitude who will now train them how to save their buddies' lives while blowing the enemy out of the zip code."

Really, it is the younger generation's comfort with and acceptance of video games that has played a critical role in their increasingly positive reception by senior military leaders. Now that the United States has been engaged in continuous warfare for over a decade, younger personnel are also beginning to replace many of the older, more technology-averse commanders. Lieutenant Colonel Michael Newell, the army's product manager for air and combat tactical trainers, explained the shift to me: "The thing about the military is that it doesn't stagnate; there's a constant generation shift. Guys who at the start of this war were colonels are now two-star or three-star generals. They were the guys that saw virtual training then, and they're now at the senior levels. You've also got captains now, company commanders, who only graduated from college four years ago; they literally have grown up with Nintendo. The junior leaders have never known a world without cell phones and video games. So they're very predisposed to being receptive to virtual training."

This emerging generation of leaders also understands that today's soldiers prefer to learn by doing — as opposed to, say, sitting in a lecture hall while their instructor takes them through a PowerPoint presentation. Just as most gamers ignore instruction manuals, opting instead to explore the game for themselves, soldiers, notes the U.S. Army Research Institute, "want to learn Army digital systems the same way that they have acquired much of their non-military digital expertise: by exploring the software and equipment to solve real problems."

What Can Video Games Do?

Despite the Pentagon's heavy reliance on video games and simulations, the military is still in the beginning stages of researching games' training and educational effectiveness. Within the services, the Office of Naval Research (ONR) is the primary outfit exploring the issue in ways that extend beyond the anecdotal and informal. According to Ray Perez, program officer in the ONR's war-fighter performance department, "We have discovered that video game players perform 10 to 20 percent higher in terms of perceptual and cognitive ability than normal people that are non-game players. [. . .] We think that these games increase your executive control, or your ability to focus and attend to stimuli in the outside world." Research on gaming is part of what Perez believes are the early stages of "a new science of learning" that integrates "neuroscience with developmental psychology, with cognitive science, and with artificial intelligence."

For the most part, academics, not military researchers, have undertaken the most serious study of whether games are effective teaching tools. Dr. Daphne Bavelier of the University of Rochester, for example, has found that people who play fast-paced, action-based video games have better visual attention skills than nongamers. These skills enable people to "focus on relevant visual information while suppressing irrelevant data." According to Bavelier's colleague Shawn Green, "At the core of these action video game-induced improvements appears to be a remarkable enhancement in the ability to flexibly and precisely control attention." For Green, the benefits of this enhancement are clear: "Those in professions that demand 'super-normal' visual attention, such as fighter pilots, . . . benefit enormously from enhanced visual attention, as their performance and lives depend on their ability to react quickly and accurately to primarily visual information."

Overall, academic research indicates that video games have at least a short-term positive effect on both basic cognitive capabilities (processing speed, visual perception skills) and higher-order thinking strategies. A literature review concludes that video games "promote dynamic cognitive activity as a player confronts challenges to be solved and obstacles to overcome that draw upon problem-solving, reasoning, and strategizing skills." This dynamic process results in the development of higher-order processes such as metacognition and justification. There is also evidence that continued game play over time modifies attention processes as well as perceptual and spatial skills. Intriguingly, improved visual skills are linked to fast-paced action games (such as first-person shooters) but not slower strategy games (such as *SimCity*). Action gamers also show improved performance in visual sensitivity, multitasking ability, and perceptual processing speed.

Research indicates that different styles of games encourage students to adopt different cognitive strategies. Students who play linear, cause-and-effect-style games adopt a strategy of finding the quickest means to an end when completing later tasks, while those who play adventure games demonstrate the ability to think proactively and infer meaning from surrounding details. In general, gamers engage in the same kinds of complex cognitive processes emphasized in school. In this way, sophisticated video games offer experiences that are consistent with play-based educational theories developed by John Dewey, Jean Piaget, and Lev Vygotsky.

Simulation games also socialize players into specific roles, which can be a highly effective way of imparting knowledge. A MacArthur Foundation report finds that games help players "develop the situated understandings, effective social practices, powerful identities, shared values, and ways of thinking that define shared communities." The report specifically cites *America's Army* as a successful example.

The MacArthur Foundation—which has spent more money researching the educational benefits of video games than any other private entity—also emphasizes the importance of "new media" literacies, which it defines as "a set of cultural competencies and social skills that young people need in the new media landscape." Almost all of the competencies and skills identified by the foundation are central to playing video games, including

- *Play*—the capacity to experiment with one's surroundings as a form of problem-solving
- *Performance*—the ability to adopt alternative identities for the purpose of improvisation and discovery
- *Simulation*—the ability to interpret and construct dynamic models of real-world processes
- *Multitasking*—the ability to scan one's environment and shift focus as needed to salient details
- *Distributed Cognition*—the ability to interact meaningfully with tools that expand mental capacities
- *Collective Intelligence*—the ability to pool knowledge and compare notes with others toward a common goal
- *Judgment*—the ability to evaluate the reliability and credibility of different information sources
- *Transmedia Navigation*—the ability to follow the flow of stories and information across multiple modalities.

As John W. Rice of the Texas Center for Educational Technology points out, a critical distinction must be made between games that "import skill and drill exercises into electronic media formats" and "cognitive Virtual Interactive Environments." The former type of game has so far dominated the educational software market. But as Rice writes, "The important distinction is that cognitive VIEs provide sufficient opportunities for complex interactions, making them suitable environments within which higher-order learning may occur." He argues that commercial titles such as *Grand Theft Auto* and *Civilization,* as well as explicitly educational games such as *America's Army,* can be considered cognitive VIEs. Among computer-based options, open-ended simulation games are viewed as having unique potential to encourage creative problem-solving and higher-order thinking.

These, as it happens, are the very kinds of games the military has most actively embraced.

The Learning Principles Behind Video Games

Probably nobody has done more to analyze and promote the learning properties of video games than James Paul Gee, the author of *What Video Games Have to Teach Us About Learning and Literacy* and a professor of literacy studies at Arizona State University. Gee argues that good video games require players to decode complex "internal design grammars," a process that emphasizes critical thinking and problem-solving skills. Game play, he writes, resembles experimental science in that it follows a cycle of "'hypothesize, probe the world, get a reaction, reflect on the results, re-probe to get better results.'" Within this process, problems are ordered in such a way "that earlier ones lead to hypotheses that work well for later, harder problems." As players strategize, explore, and take risks, they continually test, and then expand, the limits of their competence.

In his article "Good Video Games and Good Learning," Gee writes, "Some people think of learning in school .... as all about learning 'facts' that can be repeated on a written test. Decades of research, however, have shown that students taught under such a regime, though they may be able to pass tests, cannot actually apply their knowledge to solve problems or understand the conceptual lay of the land in the area they are learning." Put another way, just because a student gets an A in physics doesn't mean that she can solve a real-world problem. Schools don't necessarily have to worry about that, but the military does. By way of contrast, video games emphasize "situated meanings," in which everything that is learned is located within a specific context. This is the kind of knowledge that people retain long-term and can apply in actual practice — not just facts for the sake of learning facts, but ways of thinking and understanding that apply to consequential situations. For most people, practicing skills without a context is pointless. As Gee writes, "People learn and practice skills best when they see a set of related skills as a strategy to accomplish goals they want to accomplish." As much as anything else, video games provide a powerful, motivating context for learning and practicing new skills. Because these games are interactive, players must take an active role in this learning, making them agents of knowledge, as opposed to "passive recipients."

Another learning principle embedded in video games relates to the fact that players must take on new identities in which they are highly invested. School, Gee writes, "is often built around the 'content fetish,' the idea that an academic area" is made up merely of a "list of facts or body of information that can be tested in a standardized way." This, he says, is wrong. An academic discipline such as biology or psychology is not a body of facts but rather "the activities and ways of knowing through which such facts are generated, defended, and modified." To learn biology, one must learn to think like a biologist — to take on that particular identity, just as players do in a video game. Similarly, one learns to be a soldier not by memorizing facts about historic battles but by learning how to operate in battle. This, Gee says, is what constitutes deep learning, as opposed to the kind of learning that helps students pass tests. Video games also promote "system thinking," in which players must consider how their actions play out against the entire system of the game and the actions of everyone playing against them.

As I've already said, the emphasis on critical thinking and problem-solving skills even among the lowest-ranking personnel represents a paradigm shift for the military. Gee points out that it also parallels new theories of the postindustrial workplace, which emphasize "collaborative teams and wanting workers to actually understand their work and apply their knowledge." The difference with the military, however, is that it operates under a different incentive structure from industry and education. Unlike corporations, the military can't fire its workers if they don't succeed. Unlike schools, it can't rely on standardized test scores as indicators of success; the military has to know that its soldiers have actually learned to use the skills it has taught them.

The Other Benefits of Gaming

Cognitive and educational benefits aside, the military's embrace of video games also boils down to a more prosaic factor: money. "Live field training is very expensive in terms of time, support, ranges, fuel, ammunition, the whole gamut," notes Michael Woodman, project manager for the Marine Corps Tactical Decision-Making Simulations. Because they employ commercial off-the-shelf technology, video games are exponentially cheaper than most alternative training options. One army instructor described the benefits to me as "no overhead, no costs, no possibility of getting hurt or getting equipment damaged."

Virtual training sessions also help the military ration training grounds, which are in especially short supply today as troops return from their overseas deployments. At Fort Lewis, for example, there are seven times as many battalions as there are training grounds to accommodate them. The same goes for American bases in countries such as Germany and Korea, where, because of a lack of maneuvering space, most full-spectrum operations training is simulation-based.

Unit exercises that can be performed no more than once or twice in live training can be performed thirty to forty times on digital systems. As Lieutenant Colonel Newell told me, "When I was a company commander, for me to take a unit and go do an exercise that involved clearing two buildings on a street would have taken weeks to coordinate the range time—we would have had to do a walk-through with no weapons, no ammo, a run-through with blanks, and then finally a live-fire exercise. The whole process would take literally weeks and weeks and weeks. And it would still have some form of artificiality to it, because when you're doing live fire, there're only certain ways you're allowed to do that. With [video games], I can generate the scenario and run the soldiers through it—and the scenario is all cognitive, it's all thinking through what you are going to do, thinking through all the hazards—twenty times inside of a week." As much as anything,

Newell said, "gaming provides an ability to actually put yourself in the scenario, go through it and see it. Back up, change the scenario, go through it a different way. Back up, do it again. There are an infinite number of scenarios I can run soldiers through, because it's not about *doing* it per se, it's about having *thought* through it. So how much more experience can I give guys before we actually send them down range? When you actually get the dirt time, I can throw anything at you I want to, because you've seen it all already."

Today's military leaders are training on video games as well. For example, at the army's School for Command Preparation and the Command and General Staff College at Fort Leavenworth, Kansas, lieutenant colonels and other leaders use *UrbanSim,* a game referred to by its creator as "*SimCity* Baghdad." Developed by the Institute for Creative Technologies, *UrbanSim* focuses on counterinsurgency; players are required to manage a complex mix of civil security and control, governance, and economic and infrastructure development. Unlike *SimCity,* however, "instead of tornadoes, earthquakes, and Godzilla running around your city, it's insurgents." Securing the local population's support is crucial to success in *UrbanSim,* as is placating the host nation's leaders, not to mention running traditional military operations. The game's characters are built as autonomous agents, and they react as much to the larger environment created by the player's actions as to individual events. Colonel Todd Ebel, who directs the School for Command Preparation, says the system is like chess: "You have to think through the cause and effect of your decisions . . . you have to look two or three turns down the road."

Teaching for Real

Since 9/11, then, the military's increased emphasis on gaming has been driven by a handful of key issues: the demands of contemporary warfare; the capabilities of technologies; the changing nature of literacy;

and, crucially, economics. At the same time, the evolution of the military's post-9/11 game use has reflected the Pentagon's renewed focus on counterinsurgency.

Today, after years of efforts and advances, the military's video games have finally reached a level of complexity that matches the real-world scenarios soldiers face. What's more, years of research into the teaching properties of video games have shown how effective they can be as instructional tools. In the military's most current video games, those two elements combine to create a learning experience that is sometimes profound, as we will see.

Colonel Casey Wardynski, whom we met earlier, once gave a bracingly direct assessment of the situation to a group of educators at a conference on learning. Audience members had accused him of using video games to teach people to kill. "You know," he replied, "you should feel embarrassed that the military embraced this type of learning before you did. Our problem is that we end up with the seventeen-year-olds who failed in school, and if we teach them the way you do — that is, through skill-and-drill and standard methods — they're going to die. Because they don't learn that way. So we've got to teach them for real."

Wardynski originally came to gaming not from a learning perspective but from a recruiting perspective. Since the switch to a volunteer military in the early 1970s, the cultural connection between the services and society has largely disappeared; the military has come to be seen as "other." To attract the best recruits, the army in particular has to compete in a tight labor market for bright, technologically skilled young people who receive much of their information from online pop culture. In this environment, traditional recruiting methods, such as magazine ads, are increasingly ineffective. What's more, the image of the army propagated within this popular culture is often inaccurate or out of date.

America's Army, the video game conceived by Wardynski, is viewed as a way to redress that imbalance, albeit in a highly stylized, sanitized way. The game is a cost-effective "strategic communication tool" that not only capitalizes on a favorite pastime of the army's target audience but updates and refurbishes the image of the army held by that audience. The game's blockbuster status—from 2002 to 2008 it was one of the world's top ten online video games—illustrates the close link between the military and popular first-person shooter video games. Yet as the next chapter shows, the story of the game's development also reveals the gaps that remain in that relationship.

CHAPTER 4

America's Army: The Game

EARLY ON THE MORNING of May 22, 2002, Lieutenant Colonel Casey Wardynski woke up feeling tense in his downtown Los Angeles hotel room. As director of the U.S. Army's Office of Economic and Manpower Analysis (OEMA) at the United States Military Academy — in effect, the army's top economist — Wardynski was more comfortable in the strait-laced environment of the military than he was in the entertainment capital of the world. Yet that day, running on three hours' sleep, he faced one of the most important moments of his career: the launch of the world's first military-developed video game, *America's Army,* a project he had been directing from its inception in 1999. May 22 marked the opening of the 2002 Electronic Entertainment Expo (E3), the video game industry's enormous annual trade show, a yearly spectacle of game companies, developers, and fans convening within the glass-paneled walls of the Los Angeles Convention Center. When the center's doors swung open, *America's Army* would make its debut on the public stage. Eventually it would become one of the most popular online video games of all time, touted by the army as its most cost-effective recruitment tool ever. Lying in bed that morning, however, Wardynski felt deeply anxious about how the game would be received.

He was concerned specifically with the mainstream news media's reaction. He felt fairly certain that the games media, and the video game industry as a whole, would applaud *America's Army,* and why not? He and his team had built what they considered a first-rate game. What's more, having the U.S. Army associated with a video game could bring a degree of respectability to the industry, which had been under fire from politicians and parents' groups for years. But the major news outlets—would they pay any attention to the game? And if they *did* pay attention, what would the story be? "This is horrible"? After all, *America's Army* was intended as a tool for recruiting tech-savvy teenagers into the military; it would be available for free online, and hard copies, also free, would be distributed at army recruiting stations nationwide. The whole project was bound to be controversial.

Haunting the game's launch was the specter of the 1999 Columbine High School massacre, in which two seniors, Eric Harris and Dylan Klebold, shot and killed twelve students and one teacher. Harris and Klebold were widely reported to have been fans of first-person shooter video games like *Doom* and *Wolfenstein 3D,* and media speculation ran rampant that playing these games had desensitized them to violence and even contributed to their desire to kill. Always unpopular with parents, teachers, and policymakers, video games had after Columbine been more negatively portrayed by the news media than ever. Wardynski dreaded the possibility of the following headline greeting him in a major newspaper: "The army is using taxpayer funds to create another Columbine." The games media might write, "The army has made a version of *Grand Theft Auto*" and mean it as a compliment, but even a hint of that story angle in the mainstream media would, from Wardynski's perspective, sink the game. For political reasons, many in the Pentagon had been hostile to *America's Army* from the start, and if the top brass caught the slightest hint that the game was causing PR problems, Wardynski would never hear the end of it. He badly needed positive press to push back against the Pentagon naysayers who had told him that the game would be a disaster and who had been trying to kill the project from the start.

He needn't have worried. As he and his wife, Sue, walked the four blocks down Pico Boulevard from their hotel to the convention center, Wardynski got a message on his BlackBerry from Lori Mezoff, the game's public affairs specialist: *America's Army* had made the front page of the *Los Angeles Times.* "Is that good?" Wardynski nervously wrote back. "It's great!" Mezoff assured him. And indeed, the article made no mention of anything controversial; it simply reported that the army had developed a cutting-edge action video game as a tool for recruiting media-savvy teenagers.

Shortly afterward, as Wardynski walked around the convention center, he received news that the Associated Press and Reuters had picked up the story and that the game had made the front pages of America Online and Yahoo! What's more, he quickly determined that all the coverage was positive, as was a later report on the Los Angeles evening news. (In reality, the coverage was mostly neutral; from Wardynski's viewpoint, though, "in the news media, neutral *is* positive.") The gaming media came out strongly behind the game, just as Wardynski had predicted. One website, penny-arcade.com, even jokingly called it the "best misappropriation of tax dollars ever."

After years of effort, *America's Army* had been announced to the world, and the reaction was better than Wardynski had ever hoped for. Back in his hotel room that night, replaying the day's events in his mind, he breathed a sigh of relief and told himself, "The army can't back away from the game now."

From the moment of its debut, *America's Army* was an unqualified success. At the end of E3, eight gaming publications and websites declared it Best of Show. In the two months following the game's official launch on July 4, 2002, it was downloaded from GoArmy.com, the army's recruiting website, over two and a half million times, quickly becoming a phenomenon among hardcore gamers. *America's Army* soon became the crown jewel of army recruiting. Barely a year after the game's release, 20 percent of incoming West Point plebes reported having played it. By 2008, an MIT study noted that "30 percent of all Americans age 16 to 24 had a more positive impression of the Army because of the game and, even more amazingly, the game had more impact on recruits than all other forms of Army advertising combined."

America's Army now boasts more than 11 million registered users. The game was repurposed several years ago for use as a government training tool, and its platform is now used for dozens of training and simulation applications, including PackBot robots and nuclear, biological, and chemical reconnaissance vehicles. *America's Army* has been the focus of literally thousands of TV, online, radio, and print stories, covered in all the major media outlets in the United States. It has been a cover story in *AdWeek* magazine, and in 2009 it won Guinness World Records for "Most Downloaded War Game" and "Largest Virtual Army." Versions of *America's Army* exist for the Xbox and Xbox 360, and a variety of mobile and arcade applications are available as well. Given its overwhelming success, the army views the game—which initially cost $7.5 million to produce, about one-third of 1 percent of the army's overall marketing budget—as one of its most successful recruiting initiatives ever.

America's Army was such a hit primarily because it's what the game industry refers to as a "triple-A" first-person shooter game, meaning that it's on a par with the best commercial examples of the genre. Even if players have no particular interest in the real-world military, *America's Army* enables them to enjoy a topflight gaming experience—for free. For players who *are* interested in the United States military, *America's Army* is attractive not only for the exciting game play but also for its claim to realism; one of the game's taglines is "The most realistic army game ever!" This supposed accuracy derives in part from the spot-on depictions of weapons, uniforms, and missions in the game. The geographical backdrops in *America's Army* only add to this sense of veracity. The game's first version, released before much footage from the war in Afghanistan was available for public consumption, featured landscapes lifted directly from filmed sequences of that country, and later versions incorporate footage from Iraq and other actual theaters of war.

Also unique to the game is its focus on basic training. Before players can participate in a mission or play certain roles, such as sniper or medic, they must complete the appropriate training regimen. This requirement is coupled with the game's focus on army values. For example, a key innovation of *America's Army* is that one's own team members are always depicted as Americans, while opposing team members are always depicted as enemies. In an unusual feature for a first-person shooter game, if a player *does* shoot one of his own team members, intentionally or not, he is immediately ejected from the game in progress and thrown into a virtual version of Fort Leavenworth prison, and his player rating is reduced by several points.

Wardynski was not an obvious choice to oversee the creation of the world's first military-developed video game. Before the project began, he wasn't particularly familiar with or interested in video games. His interest was in reforming the traditional recruiting paradigm, because many recruiters, he says, because of the incentives they face, "don't care about kids—they only want kids to *sign up* for the army, they don't care whether kids *succeed* in the army." However, Wardynski was after the kinds of recruits who could succeed and even spend their entire careers in the military. As such, he thought the game should require investment before it yielded reward—specifically, basic training as the investment required for the reward of hitting the battlefield. (The thought that war should be a reward did not trouble him.) He felt that the motivating question guiding the design of *America's Army* should therefore be, why does the army have basic training? He wanted to emphasize the army's "seven core values," which are listed on GoArmy.com: loyalty, duty, respect, selfless service, honor, integrity, and personal courage. By focusing on basic training and army values, Wardynski thought he had found the right spin for justifying the game to the army's leadership and to anti–video game crusaders such as Senators Joe Lieberman and Herb Kohl, who had headed a joint congressional investigation into the industry's marketing of violent video games to minors.

In person, Wardynski comes across as simultaneously deferential and dominant, his voice gentle but insistent. Though he was raised in the suburbs of Chicago, his voice has a tinge of a southern accent. He is handsome and fit, of medium height and build, and his piercing gray-blue eyes are his most striking feature; when he is talking, his gaze remains locked on his listener. A single question is likely to send him spinning off on a half-hour spiel, one in which his sharp intelligence is readily apparent, as is his ability to guide any conversation, no matter the topic, back to the issue *he* wants to discuss. He is bracingly blunt, fiery in his opinions, and withering in his criticisms of the Pentagon. Yet few would deny that the strength of Wardynski's personality, along with his intelligence, is what guided *America's Army* to its ultimate success.

People can see what they want to see in *America's Army*—it can properly be regarded as both teen-focused propaganda that militarizes young minds and a striking example of military innovation. It stems from and is a tool for war, and it represents a direct intrusion of the army's hand into the homes (and minds) of children and teenagers. But as a government contracting model, it also represents a streamlined (and exponentially cheaper) improvement over decades of Pentagon waste. The game's ability to transcend a simple narrative is part of what made it an object of interest and controversy from the start.

Ultimately, the idea of using a game to brand an enterprise, to tell people what that enterprise is like as a way to motivate them to either respect it or join it, is one of this game's legacies. The success of *America's Army* helped popularize the idea that a game could promote a brand. In this way, the game has been as influential in the world of marketing as it has been in the military.

A Cocktail Party in Calabasas

The concept for *America's Army* arose in 1999, when the army's recruitment numbers were at a thirty-year low. In that year alone, the army missed its target number by seven thousand — the worst showing in the history of the all-volunteer force. In response, the RAND Corporation, a federally funded public-policy think tank, began conducting research into why teens were not attracted to the army. At the time, Wardynski was working on his army-funded doctorate in policy analysis at RAND's Pardee Graduate School, and he talked regularly with the lead researcher on the project. RAND's data indicated that fifteen- to twenty-four-year-old males no longer particularly valued the skills and values historically associated with the army: leadership, discipline, patriotism, doing something for the community, traveling the world. Instead, they wanted to develop the skills valued by future employers, the skills needed to "succeed" in life, to do something mentally challenging; they wanted to enjoy their lives, and they wanted to get paid well right away or else learn the things that would help them find a well-paying job later on. None of these interests matched what young males thought the army could provide. Indeed, the only real motivation for joining the army was receiving tuition assistance for college. What's more, there were nearly as many reasons *not* to join the army as there were to join: the loss of personal freedom, the irrelevance of military service to future careers, the possibility of hating the army once one enlisted, the loss of a "normal" lifestyle. (Tellingly, in the post–Cold War nineties, concerns about the physical dangers of military service ranked far below lifestyle concerns.)

The issue of military service itself wasn't the only problem. Equally important, teens rated the army as by far the least desirable of the four military services. The army is low-tech, they said; it's sweaty, dirty, and dangerous, and it's for people who have no other options in life. On a PowerPoint slide divided into quadrants, the RAND researchers mapped out the army versus the navy, air force, and Marines. The worst quadrant to be in was southeast, which indicated low-tech and ordinary. According to respondents, the air force was high-tech and elite; the Marines were elite and sweaty but not low-tech (though this is not true—the Marines *are* comparatively low-tech); and the navy occupied a middle ground. The army, located squarely in the southeast, owned the worst piece of turf.

In an effort to reverse this negative trend, the secretary of the army, General Louis Caldera, ordered four working groups of senior officers and civilians to come up with innovative ways to boost enlistment. Adopting the language of corporate America, Caldera wrote, "Our [selling] strategy must be based on solid market research and our messages must be better targeted to those segments of the market where we envision increasing our market share." Booz Allen Hamilton, a major defense consulting firm, was brought in to organize and evaluate the various alternatives for new recruitment initiatives. The firm's Working Group B portfolio of proposals ranged from the mundane to the ridiculous: perhaps personal computers could be given to new enlistees, or recruiting stations could be relocated to areas with greater foot traffic. Perhaps teenagers could be offered laser eye surgery to join the army; perhaps the jails could be emptied; perhaps Ethiopians and Somalis and other foreigners could be recruited to beef up the army's numbers. Patrick Henry, assistant secretary of the army for manpower and reserve affairs, stated that young Americans needed to be convinced that the army was not "an employer of last resort." To this end, he said, the army had to "find a way to link together emerging technologies and a soldier's drive toward self-edification." As part of this effort, the Pentagon raised its recruitment budget to a record-breaking $2 billion a year.

As it happened, in January 1999, a few months before Army Secretary Caldera issued his call for new recruitment strategies, Wardynski attended a cocktail party in the affluent city of Calabasas, California, next to Malibu in the Santa Monica Hills. Calabasas houses a number of technology companies that dot the road next to the 101 freeway, an area known as the "101 technology corridor." From 1997 to 1999,

Wardynski, his wife, and their three children lived nearby in a rented house while he finished his graduate course work at RAND.

On the warm winter night of the cocktail party, Wardynski ended up next to the party's fire pit, drinking margaritas and chatting with a man named Jesse, a friend of a friend. Jesse was describing his business, which involved distributing media on CDs for the movie industry and assorted advertisers. Whatever the content—movies, commercials, instructional videos—Jesse would dump it onto a CD and ship it off to a waiting customer. Intrigued, Wardynski asked Jesse about his distribution costs. The answer surprised him: it cost little more than a dollar to ship a CD to its destination. As an economist—and because recruiting was a key focus of the Office of Economic and Manpower Analysis—Wardynski immediately pricked up his ears. He knew what the army was spending to put its message into homes, and this number was breathtakingly low in comparison.

As he continued talking with Jesse, Wardynski realized that a CD could hold a great deal of content. Even better, a CD could be shipped directly to people's homes. But what exactly, Wardynski asked himself, would the army's content be? Soon after his discussion with Jesse, he did some research on the Internet and discovered that the computer game company NovaLogic was a quarter mile from his house. The company, it turned out, had just provided training software for the U.S. Army's Land Warrior, a program designed to integrate commercial, off-the-shelf technologies into a complete "soldier system," meaning that each soldier would become "an individual, complete weapons system," equipped with "weapon, integrated helmet assembly, protective clothing and individual equipment, computer/radio, and software." On top of that, the company had recently produced *Delta Force,* a popular first-person shooter video game.

Wardynski called NovaLogic, told them who he was, and requested a meeting to discuss an idea that was just then taking shape in his mind. He had noticed that when he took his children to Best Buy, they would hang out in the video game aisle, and it would take a great deal of effort to drag them away at the end of the visit. At home, his children and their friends seemed obsessed with military-themed video games, which they played in every spare moment. On a later trip to Best Buy, Wardynski decided to look more closely at the video game aisle, and what he found there surprised him: roughly two-thirds of the games were about armies. Some of them were Roman armies, some of them were futuristic armies, but they were armies nonetheless. This finding, coming on the heels of his talk with Jesse, compelled Wardynski to call NovaLogic.

The company was happy to arrange a meeting. In March 1999, Wardynski sat down with both the president and the CFO to discuss whether it was possible to make a computer game about soldiering—not about fighting, but about what it took to *be* a soldier—and if so, what the disk space and cost requirements would be. They talked about *Delta Force* and about the cost of doing a video game. Again, the pricing is what drew Wardynski's attention: creating a game cost between $1 million and $2 million—cheaper than making TV commercials. What's more, a video game could be a far more compelling medium for delivering information than a thirty-second television ad.

"Let me get this straight," Wardynski asked the CFO. "We could have *Delta Force II*, your top-tier game, for one or two million dollars?"

"No," the CFO responded. "*Delta Force II* would be far more expensive. We'd have to give you one of our old games."

"How is the army going to appear high-tech and cutting edge if we start out with a product that's no longer salable?" Wardynski asked himself. Nonetheless, he found the talk with NovaLogic useful: he learned how games are built, the number of staff and length of time required, the technology involved. The discussion with NovaLogic helped crystallize key aspects of the game that Wardynski was starting to envision: online, multiplayer, soldier-focused, small team–based. Primarily, the discussion convinced Wardynski that building a game for the army could be done. The question that remained for him was, how do you do it?

The Economic Rationale

Wardynski's radical notion of how a video game could revive the army's hidebound recruiting process stemmed from his background as an economist. He felt that young adults were getting the bulk of their information from computers and the Internet and that the army, to attract recruits, needed to adapt to the popular culture of the information age. The video game, Wardynski thought, could be a highly effective way to do so, primarily because it could be delivered directly into the homes of its target audience. He says that *America's Army* was explicitly designed to target twelve- and thirteen-year-old boys — in his words, "to capture youth mind-share" — who had yet to decide what to do with their lives. "When a kid starts thinking about what he's going to do with his life, it's not at age seventeen, it's more like age thirteen. You can't wait until they're seventeen," he told me, "because by then they will have decided that they're going to college or to a trade school, or they'll already have a job that they're planning to stay in. You have to get to them before they've made those decisions."

By connecting directly with young males, a video game could offset what Wardynski calls the "market failures" that had led to such a steep drop in army recruitment numbers at the end of the 1990s. The idea was less about entertainment than about economic theory. The shift from a draft-based army to an all-volunteer force in 1972 had gradually pushed the army away from mainstream American culture, while the popularity of the Internet and the increasing ubiquity of computers in middle-class teenagers' lives had rendered the army's traditional advertising and recruiting strategies ineffective.

When the draft was in place, the army didn't need to get people excited about joining up, but the advent of an all-volunteer force meant that people had to be sold on soldiering as a career choice. Recruiting became a two-way interaction, one that had important repercussions for the army. "Now I'm trying to match your *interests* to some piece of this organization, where before it was like we're going to match your *abilities* to some piece of this organization," Wardynski says. "That's a whole different approach. And that interest piece is a killer, because we may have jobs that aren't interesting, or they're not interesting at the price we're willing to pay, and so there's an ongoing discussion and conflict between what we're willing to pay, what this job is worth in an economic sense."

Economists tend to approach market analysis as if people have perfect information and make economic choices rationally. But according to Amos Tversky and Daniel Kahneman, two seminal figures in behavioral economics, this approach ignores the facts that information can be expensive to acquire and assimilate, that there are innate biases in human decision-making, and that instead of being rational, people tend to base their decisions on heuristics, or decision-making shortcuts. These heuristics can lead them to be overly influenced by the vividness and availability of information in their immediate environment. Thus they make decisions based on imperfect information. Behavioral economists believe that first impressions are crucial, because it takes a lot more information to change people's minds than it does for them to make up their minds in the first place. In Wardynski's reading, this is the result of human evolution. "Genetically, the people who would end up surviving were the ones who had embedded shortcuts," he says. "The shortcuts are: the first information you get is the most important information you get, and vivid information is crucial. If you're a caveman running around, hunting for dinner with your buddies, and some big furry cougar jumps up with claws and teeth and devours your friends, that's real vivid. It's important to you because you could be next. So the people who could figure out real quick from vivid information what to do with it were the ones that survived and contributed to the gene pool."

A great deal about the army is vivid; that in and of itself was not the issue with recruiting. Rather, the problem was (and is) the particular type of vivid. The army, Wardynski says, is primarily "vivid-dangerous, vivid-bad, vivid-abuse, vivid-degrading—it sucks, right? *Platoon,* all the movies made about Vietnam and about how shitty the army was and the cruddy leadership and the lousy conditions. Or vivid-heroic, but in a negative way. *Saving Private Ryan:* heroic as hell, but I don't want to be there!" Wardynski believes that army veterans can also do more harm than good in this regard. Even if they love the army, he says, the stories that veterans tell are going to be vivid because "nobody wants to listen to some boring story about hanging around at the forward operating base, eating steak and lobster served up by Brown and Root in an air-conditioned mess hall in Fallujah. They want to hear about the ambush or the getting blown up or how you overcame the screwed-up army. So even if veterans love the army, the stories they're going to tell you are about somebody getting blown up. So from a recruiting perspective, they're not our friends — we *have* no friends! We're not even friends with ourselves, because we don't know how to talk about ourselves, or when to talk about ourselves, or whom to talk to."

Wardynski felt that the army had fallen behind on two fronts: first, in making up-to-date information about itself available, and second, in assimilating the millions of strands of information about the army that *did* exist on the Web. This, then, is where Wardynski thought a video game could be useful: it would be located in pop culture, where young people could find it, which would take care of the information search problem. Because the game would feature an immersive, engaging format, the assimilation costs would go down, too. Both of these things would chip away at the market failure problem.

Another important issue for Wardynski was disintermediation, a term in economics that refers to cutting out the middleman. As opposed to going through traditional distribution channels, a company disintermediates by dealing directly with its customers, often through the Internet. "You can ask yourself, 'What do I know about the army, and where did I learn it?' TV, print, movies, the news — those channels entail intermediation, and they're intermediating our story," Wardynski says. At the time, he felt that what most Americans knew about the army was completely out of date. Soldiers in the post-Vietnam volunteer army had a much better quality of life than those in the Vietnam-era draft army, and yet people didn't know this. (For example, in the draft army soldiers needed to get permission from their unit commanders in order to get married, whereas in the volunteer army they don't.) The army remained, in effect, locked in a time warp. If young people watched TV, they thought soldiers were still "living in Gomer Pyle barracks," Wardynski says. They thought soldiers couldn't date, that they couldn't own a car. A video game would help the army to disintermediate these outdated images by delivering large volumes of its own content directly to young people's computers.

Paradoxically enough, Wardynski also wanted the game to deter people who might not be well suited to military service from joining the army. His reasoning here was again economic: the army was losing $400 million a year on recruits who dropped out of basic training because the military was not what they had expected. The army has a dropout rate as high as 18 percent; given that the training costs can be more than $100,000 per soldier, that represents a serious financial loss. By letting young people "test-drive" both basic training and actual battle, Wardysnki saw *America's Army* as a way to weed out those who might drop out later at vastly greater expense to the government.

He rejected any criticism that a market-based focus in relation to young adults and military service was a cause for concern. "There're plenty of people that would say, 'Well, aren't you targeting young people?' Well, who would you *have* us talk to? Old people? I mean, what army would *that* look like? Of course we're talking to young people, and of course we're talking about being in the army, but we want it to be a good fit."

Finding a Sponsor

Following his visit to NovaLogic in early 1999, Wardynski began to pursue the question of how the army could get access to a top-tier game without having to pay for an entire professional game studio. Though he held a prominent position in the army, his Office of Economic and Manpower Analysis did not have the purview or the budget to build a video game on its own. He needed a sponsor.

At the time, Wardynski's mentor and boss was Lieutenant General David Ohle, the army's deputy chief of staff for personnel. During the months that the video game idea was crystallizing, Wardynski was working on another Pentagon project for Ohle about the financial impact of being married to a soldier. (This was the topic of Wardynski's dissertation.) Though this project took most of his time, the game idea continued to excite him. Finally he decided to take action. He wrote an e-mail to Ohle's point person for recruiting, Major Keith Hattes, outlining his video game idea and asking for Hattes's assistance.

Hattes had recently graduated from the Naval Postgraduate School in Monterey. When he received Wardynski's e-mail, he wrote to several faculty members at the school, saying that he wanted to "canvass [their] interest" in helping the army's recruiting efforts by developing a new "technological initiative" proposed by the director of the Office of Economic and Manpower Analysis at West Point. Lifting from Wardynski's e-mail, Hattes wrote that the idea was "to build a game-oriented, virtual reality-based, distributed (web-based/linked), interactive, and adaptive simulation that would allow potential recruits to explore virtual Army adventures and progress through levels of expertise and areas of specialization over time based on their abilities and interests." The goal, Hattes noted, was to mass-market the simulation by CD or online "to targeted population pools to assist in expanding market audiences and also pre-screen potential recruits."

In proposing the development of a computer-based video game/simulation, Hattes and Wardynski were adding to the long legacy of military research and expertise. In the years immediately preceding Wardynski's proposal, no one had worked harder to bring the worlds of the military and the entertainment industry together than professor Mike Zyda, who was one of the ten recipients of Hattes's e-mail.

A Clash of Cultures

After months of meetings—and months of navigating the internal politics of the Pentagon—Wardynski convinced his superiors that the game should be built by Zyda's brand-new MOVES Institute at the Naval Postgraduate School. In May 2000, the secretary of the army, Louis Caldera, approved the Army Game Project as a recruiting initiative, one of many that the army was taking on at the time. Caldera assigned the game project to the assistant secretary of the army for manpower and reserve affairs, Patrick Henry, who in turn made John McLaurin, the deputy assistant secretary of the army for military manpower, the project's executive director. Though his actual involvement was only occasional, McLaurin, in name, oversaw the Army Game Project for the first several years of its existence.

Technically, a project like *America's Army* should have gone to STRICOM, the army's simulation and video game office, and been funded through the army's research, development, testing, and evaluation budget. Instead, with McLaurin's blessing, Wardynski ran the project out of OEMA, with funding provided by the secretary of the army's VIRS account, which goes toward recruiting initiatives that fall outside the normal channels. VIRS money is what's left over at the end of every year after the funding for traditional recruiting initiatives (television, advertisements, funding for recruiting stations) is taken out. Wardynski valued the relative autonomy this arrangement allowed, but he was also aware of its tenuous nature: not only does VIRS money fluctuate on a yearly basis, its release to specific projects depends wholly on those projects' so-called godfathers. In the case of *America's Army,* that godfather was McLaurin. Wardynski knew well that if McLaurin disappeared, so did the money, and thus the project.

Still, on a practical, day-by-day level, *America's Army* was Wardynski's baby, and he continued to insist that *America's Army*—at that point called *TAG,* as in "the army game"—center on basic training and army values. In this he was following a well-established tradition of addressing—and enhancing—the morals of its soldiers. As we've seen, George Washington himself sought to improve the morals as well as the literacy of his charges by instituting a program of Bible study at Valley Forge. This deeply ingrained thread of moral instruction achieved a new prominence with the onset of World War I in 1914, when calls for universal military training in public high schools and colleges became more vocal. Military education, like public school education in general, was viewed by its promoters as a means for resolving a welter of social issues. As scholars Lesley Bartlett and Elizabeth Lutz write, this included eliminating a "'moral rot' that had become symbolically associated with the country's growing wealth; the lack of a sense of duty or loyalty in the massive numbers of new immigrants; and the social disorder of strikes and other labor unrest." Military training was to inculcate discipline and respect for authority, thereby developing "the 'moral qualities' of 'good citizenship'" and a "quickened patriotism."

Attributing a quasi-religious role of moral betterment to the military continued through the Truman Doctrine and the Marshall Plan, only to be seriously damaged by the experience of the Vietnam War. The military spent decades struggling to recapture its pre-Vietnam image, a goal that the terrorist attacks of 9/11 seemed finally to place within reach. In keeping with this goal, Wardynski saw *America's Army* as an ideal tool for rebuilding the military's public image.

As part of Wardynski's efforts to focus the game on army values, he created a Red Team in the Department of Social Sciences at West Point, where OEMA was located. (In the military, a Red Team is an opposing force.) Headed by the chair of West Point's political science program and featuring a combination of political science, international relations, and communications professors, the Red Team's job was to review the existing antigames literature in order to figure out the areas in which *America's Army* would be most vulnerable to criticism. By anticipating potential criticisms in advance, Wardynski hoped to guide the game's development in such a way that these concerns could be avoided, or at least minimized.

As the Red Team searched LexisNexis, the names of two antigames activists kept popping up: Lieutenant Dave Grossman, author of the best-selling book *On Killing: The Psychological Cost of Learning to Kill in War and Society,* and Jack Thompson, a Florida lawyer whose endless crusades and media-driven lawsuits had made him the bane of the video game industry. Grossman's primary argument was that video games and simulations desensitized soldiers to the act of killing. Thompson's approach was to sue game companies on behalf of parents who had lost their children in school shootings.

With Grossman and Thompson in the back of their minds, Wardynski, OEMA, and the Red Team developed the design criteria — the Ten Commandments, as they thought of them — for *America's Army.* They knew their target audience of male teenagers would be interested primarily in the game's combat elements, but they recognized that these elements could not be the game's public selling point. Therefore, they decided to emphasize that the army focused on the *sanctioned,* not random, use of violence. "The army, job one is to fight the nation's wars on land," Wardynski says. "And that entails the managed use of violence. And we use as much violence as it takes to do the job. And we put that story in the game — that is, the army's story." This is what Wardynski saw as *America's Army*'s key attribute: the violence in the game would be backed by America's "credibility and reputation."

Along with the sanctioned use of violence, OEMA's primary criterion for the game was that it take into account "international sensitivities." Because of this, OEMA initially struggled to design the game's bad guys. "That was pretty much solved for us on September eleventh," Wardynski recalls, but there was still the challenge of how to depict al-Qaeda. "It could've been stereotypically Arab-looking guys," he says, "but we chose not to do that, because al-Qaeda isn't stereotypically Arab-looking — they have Swedes, they have Somalis. So we made the bad guys look like everybody."

A related issue was who would play the enemy in the game. "If you could be the enemy," Wardynski realized, "then we couldn't bind you to the rules and standards of conduct that we wanted." OEMA and the Red Team therefore decided that users would play only as friendly forces. In a move unprecedented in commercial gaming, *America's Army* used mirror imaging, so that players would always see themselves and their team as American, while the game's other users would appear to be from the opposing force.

"We need the focus on basic training and army values to keep the politicians and generals off our back," Wardynski told Zyda. "The army isn't in the business of having first-person shooter games." Here a fundamental conflict arose between Wardynski's and MOVES' visions for the game: from a gamer's perspective, basic training is boring. Values, in the context of a video game, are boring. The people at MOVES were all hardcore gamers, and they felt strongly that Wardynski was wrong: action needed to come before values, or no one would want to play the game. Zyda had hired Mike Capps, a former MIT graduate student with years of experience in computer graphics and virtual reality, to head his development team. As Capps saw it, the MOVES team needed to build a game as good as *Counterstrike* or *Half-Life* or *Unreal Tournament*—a game, in other words, that was not only high quality but also fun—and then insert the army's message underneath that. As Capps pointed out to Wardynski, "If we follow the path of, 'Let's plan out the army's message first and then try to make that fun,' the game will fail."

In addition to Mike Capps, Zyda's team included creative director Alex Mayberry, who had spent nearly a decade in the commercial game industry. The army supplied Lieutenant Colonel George Juntiff as a design consultant; Zyda credits Juntiff with essential service in making the game look and feel as army-authentic as possible. Zyda's MOVES Institute, meanwhile, provided a steady stream of graduate students conducting research in everything from streamlined graphics algorithms to analysis of the psychological dynamics of immersion. This cutting-edge research was applied repeatedly in the first two years of the game's development.

Capps selected Epic Games' highly regarded Unreal Engine as the technology on which the team would build its army modifications—

that is, the game's content. For $500,000, the team purchased a license for the game engine via the Naval Postgraduate School's contracting office. Because Unreal Engine was recognized within the world of gaming as a topflight technology, both MOVES and Wardynski thought it would lend credibility to the army game.

As the developers and Wardynski hashed out the game's details, the developers worked to strike a balance between army authenticity and gamer enjoyment. For example, one mission involved taking a radio tower out of commission. In real life, Rangers might simply blow up the tower—a scenario that wouldn't suit the game because it would happen so fast. Instead, the developers designed the mission so that players need to identify friendly forces, battle enemy insurgents, and protect NGO workers until communications specialists manage to overwhelm the tower.

Wardynski rejected several of the developers' proposed missions owing to their lack of military realism, while the developers discarded several of Wardynski's desired effects for technological reasons. For example, the game contains a parachute jump, but the jump's intended beach landing had to be scrapped because accurately rendering water requires a great deal of hardware. The use of rope in the game was abandoned for similar reasons.

By May 2000, Wardynski and the developers had agreed on ten initial levels for the game. Led by Lieutenant Colonel Juntiff, the MOVES team immediately began visiting army posts across the country—they eventually traveled to nineteen in all—to gather data. They visited a rifle range at Fort Polk; they photographed weapons at Fort Lewis; they observed house-clearing operations at Fort Benning. Everything that would appear in the game was either photographed or shot on video. Thousands of relevant sound effects were recorded. The braver developers even engaged in a tower jump and were willingly attacked by dogs. In one memorable episode, the MOVES team participated in a late-night Black Hawk helicopter ride over a barrage of live shells pinging the ground below. Not only did these experiences provide fodder for the game, they gave developers the vicarious thrill of engaging in real military action. This marked a vivid departure from the developers' normal routine of spending seventy-hour weeks ensconced in front of their screens.

To ensure the fidelity of the game's animations, the developers outfitted soldiers with motion-capture sensors. The soldiers were then filmed enacting various operations — say, throwing a grenade — in accordance with military procedure. (The results were so accurate that they've been used to train West Pointers.) Not all of this translated directly into the game. The developers felt that if characters' running, for example, accurately reflected humans' true speed, the relative slowness would drive gamers away. In several instances, then, the animators cropped or otherwise streamlined the action so that gamers would remain engaged. To do this, they worked frame by frame through their videos to pinpoint key moments that the eye needed to observe; all intermediate movements were removed. When speeded up, the resulting effect resembled that of a flipbook, with the eye witnessing the motion as a continuous scene.

Russell Shilling, an associate professor in the Naval Postgraduate School's Operations Research and Systems Engineering Departments, led the MOVES team's sound design efforts. He had spent years consulting with Hollywood's leading audio designers and engineers, and his research at MOVES focused on auditory psychophysics. Inspired by his experience, Shilling wanted the game to feature the most evocative, intricate sound yet heard in a first-person shooter.

To determine how best to achieve this, Shilling and his graduate students embarked on an ambitious research agenda. They wanted to learn more about how sound evokes presence and emotion in a video game and whether this improves the way players execute memory-related functions. To find out, they employed a variety of methods, including measuring how a player's heart rate and breathing changed in response to certain auditory cues.

Shilling and his students also scoured professional libraries for relevant sound effects. When a weapon is fired in the game, for example, the impact of the bullet — or, in some cases, the grenade — is accompanied by a complex, multilayered sound: shell casings clatter off walls; bullets roar past the ear; glass explodes and shatters on the floor. Tinnitus accompanies the aftermath of a grenade.

The MOVES team also took pains to make the game's physics significantly more accurate than those in most popular commercial shooters of the time. In *America's Army,* a rifle may gently swing, depending on the avatar's breathing; moreover, there is a notable kickback when the rifle fires. (Sometimes weapons even jam.) Depending on the caliber and type of weapon, after a bullet is released, that bullet may penetrate deeply into wood, adobe, or dirt, or it may ricochet off a steel surface — whatever it would do in real life. The avatar's distance from the target is also taken into account.

Zyda and his team intended these physics to affect the way players make decisions in the game. For example, players quickly learn that if they set off flashbang grenades at close range, they temporarily go blind and deaf. If they stay too close to walls while moving, they face an increased risk of being nailed by ricocheting bullets. If they run around firing their weapons quickly, their scores are much lower than if they stay stable and take time to point and shoot accurately. To drive this last point home, players are penalized for causing friendly-fire incidents.

Despite the MOVES team's scrupulous efforts, conflict continued to mark the relationship between the developers and Wardynski during the two years of the game's construction. Again, much of it came down to Wardynski's insistence that the game's design criteria focus on army values. "You can't just have a billboard in back of the game listing those values and call it done," he told Zyda and Capps. "Somehow you've got to *show* those values and why they're important." Wardynski gave them examples of the kinds of missions he wanted in the game: "Have a basic training scenario where players can choose to do the wrong thing and then show them *why* it's the wrong thing." As an example, Wardynski told the team, say that players want to be a medic in the game, and when they're taking the medic exam in basic training, they decide to cheat and look at another player's paper. When they're on the battlefield and they're responsible for helping wounded soldiers, the fact that they cheated in basic training means they don't have the skills to do their job, and other players will know they're incompetent and not to be trusted.

Wardynski also wanted the game to show inherent conflicts in the warrior ethos, where one element says, "Don't leave a fallen comrade" and another says, "Mission first." He instructed the MOVES team to develop a scenario in which players might have to leave a wounded or fallen comrade in order to accomplish their mission. "These are the kinds of things that make us an American army," he told Zyda and Capps. "That's what will make this game special. You're not going to see that kind of thing in *Ghost Recon* or *Halo.*"

The back-and-forth between Wardynski and the MOVES team persisted after the game's initial release. Wardynski wanted to add additional nonkinetic (that is, non-battle-related) scenarios to the game, because he felt that the developers still hadn't focused enough on army values. He continued to worry that *America's Army* would be labeled solely as a first-person shooter. To avoid this, he wanted the next scenario to be Airborne School, where players would learn to be army paratroopers, which would be fun as well as nonkinetic. The MOVES team, however, wanted the next scenario to be Sniper School, because that would appeal more to teenage boys. Wardynski felt the word *sniper* had a negative connotation in the civilian world, conjuring up images of a man in a bell tower shooting at innocent people. In the army, a sniper is simply an advanced marksman, but Wardynski felt that this distinction would be lost on the media and the public. The developers insisted that calling the scenario Sniper School would make for better game play. Wardynski finally compromised, saying, "Fine, we'll have the scenario, but it will be called Advanced Marksman, and we'll only do it after we do Airborne School."

The developers eventually had their revenge: as soon as Advanced Marksman was released, they labeled it Sniper School in their discussions on the game's website, a name the community of players quickly adopted as their own.

Playing the Game

By now there are several versions of *America's Army*. Let's take a closer look at perhaps the most popular one, *America's Army: True Soldiers*, which was developed for the Xbox. Like the game's other versions, *True Soldiers* is a first-person shooter with a nine-part individual training section and a subsequent nine-part team section. These exercises, which take hours to play, lead up to the multiplayer function on Xbox Live.

The actual game play resembles that of most first-person shooters — you shoot, you maneuver, you find re-ups (resupply packs, medic kits). The interface is also fairly traditional — an indicator in the bottom left corner of the screen shows your stamina level, ammunition supply, and so forth, and there is a map in the bottom right corner. In contrast to most first-person shooters, however, *America's Army* is surprisingly text-heavy. In addition to the instructions, the manual features a large amount of text addressing players as potential recruits. Take the following paragraphs, all of which follow from Wardynski's vision for the game:

> On June 14, 1775, the Continental Congress created the Continental Army, which, after the Revolutionary War, became the United States Army. The U.S. Army and the millions of Soldiers who have served over the past two centuries have been guided by the core values of Loyalty, Duty, Respect, Selfless Service, Honor, Integrity, and Personal Courage. These values have been the bedrock of the most critical component of the armed forces, responsible for defending U.S. interests worldwide and for winning the nation's wars.
>
> As the largest of the four armed services in terms of personnel, state-of-the-art equipment, infrastructure, and training facilities, the U.S. Army brings talented Soldiers together with cutting-edge technology and training to produce the most capable and respected military organization in the world. The Army . . . operates more aircraft than the U.S. Air Force, owns more water vessels than the U.S. Navy, and has conducted more amphibious operations than the

> U.S. Marine Corps. More than 90% of the nation's Special Operations Forces are in the U.S. Army. An Army so important, so large, and so high-tech invites the best and the brightest young people to strengthen their minds, bodies, and character through service to America.
>
> America's Army: True Soldiers is the only official U.S. Army game on Xbox 360 and was built and tested with U.S. Army Soldiers at every level of production. No game can fully create the reality of serving in the U.S Army, but America's Army: True Soldiers provides authentic insights into life and success as a Soldier . . . Throughout this experience, you will grow and improve as a person and a virtual Soldier. You will have the opportunity to earn Respect and gain Honor while serving as one of America's Army: True Soldiers. Prepare for the challenge of this experience. Prepare to be a True Soldier.

The game opens with the *America's Army* logo paired with the sound of reveille. Beneath it, the sentences "Empower yourself. Defend freedom" appear to the sound of a shotgun being cocked. This is followed by the menu screen; click on Basic Combat Training and you are shown a scrolling list with a short description of each training mission, including the following:

> M9 Training
> *Staff Sergeant Johnson will brief you on the operation and function of the M9 pistol. You will then practice with the weapon and then be tested for basic qualification.*
>
> M24 Training
> *Staff Sergeant Johnson will brief you on the functionality of the M24 Sniper Rifle. You will then practice with the weapon and then be tested for basic qualification.*
>
> M203 Training
> *Staff Sergeant Hernandez will brief you on the functionality of the M203 Grenade Launcher. You will then practice with the weapon and then be tested for basic qualification.*

M136 AT4 Training
Staff Sergeant Hernandez will brief you on the functionality of the M249 Squad Automatic Weapon. You will then practice with the weapon and then be tested for basic qualification.

Obstacle Course
Learn individual movement techniques (IMT) to combine moving with shooting.

MOUT
Learn the basics of team and squad movement on the Military Operations on Urban Terrain training course.

All but the final two basic training exercises follow the same template: You are at an outdoor firing range surrounded by low hills and scrubby vegetation. You can hear birds chirping and the sound of rounds being fired at distant ranges. A sign reads military firing range: keep out.

Standing in a sandy clearing, one of three ethnically various but uniformly brusque sergeants introduces a weapon, explains in detail how it is used in the army, and instructs you on its in-game use. You cannot move or engage in any activity when the sergeant is talking, and you cannot skip his speech even if you've already heard it. The vocals and the visuals are slightly out of sync.

Following the sergeant's introduction, you are directed to pick up your weapon from a table nearby and proceed to one of two firing ranges. In one range, you can practice shooting at targets, typically little green men who pop up and down at random. In another, you are tested on your rifle skills and knowledge. You cannot proceed to the next training level without passing this level's test.

If you wander away from the shooting range, your sergeant will bark, "Return to your training yard!" or "I gave you an order!" If you ignore the sergeant and continue exploring, you fail the mission, which immediately ends it, and lose Honor points, which lowers your overall score. For the most part, if you point your gun at your commanding officer, it will disappear from your hands. If you do manage to shoot him, however, you will fail the mission and lose Honor points.

The final two training exercises, Obstacle Course and MOUT, break from the pattern of the previous ones. In the obstacle course, you run from station to station toward your supervisor, clambering over logs, sprinting over a rope bridge, maneuvering through a shoothouse (a live-fire, 360-degree training range), administering first aid to a colleague, crawling under low wires, and shooting various weapons.

The MOUT is the final Basic Combat Training level, and the first level in which you play as part of a team. You run through an abandoned town with your team, shooting various enemies, who are marked as bad guys with red marks over their heads, in typical video game fashion. Since this is a training exercise and doesn't use live fire, the men who've been hit (they're all men) sit on the ground with a sulky look. The objective is to follow your sergeant to a bell tower, where you shoot a sniper and defend the structure against oncoming enemies.

Once you have completed Basic Combat Training, you move on to the war games, of which there are eight. Each is designed to familiarize players with the roles of rifleman, automatic rifleman, sniper, and grenadier. Game play for each mission is capped at thirty minutes.

The first mission, Operation Gray Wolf, takes place in a small village crisscrossed by dirt paths. You are in a team with four members, led by a sergeant, and your goal is to shoot enemies (again, no one dies; people just sit down) manning several machine-gun checkpoints in the area. There are no civilians.

At various points in the mission, your sergeant will bark warnings, encouragement, or directions: "Don't get cocky"; "Not bad, not bad at all"; "Something about this just doesn't feel right"; "This is a soup sandwich"; "All right, everybody into position"; "Keep your eyes open for enemy activity"; "Find your area of responsibility"; "Take that target out."

If you are shot, you must sit down; you cannot move until a teammate gives you first aid. If you and all your teammates "die," you fail the mission and have to start again.

Aside from its focus on basic training and its discouragement of friendly fire, what most differentiates *America's Army: True Soldiers* from its commercial rivals is its Extras feature, which is intended for recruiting. It includes sections on the army core values; a week-by-week description of the army's basic training regimen; a set of enlistment requirements; a list of soldier and officer ranks; a collection of army trivia ("Soldiers can earn up to $72,900 to help pay for college"); full descriptions of dozens of jobs available in the army; descriptions of various weapons, including information on ammunition type, magazine capacity, and rate of fire; and twenty-eight video profiles of people serving in or affiliated with the army, including a sniper, a pilot, a physician, and an "army mom."

Diversifying the Product

In the two years following the release of the first *America's Army* in July 2002, relations between OEMA and MOVES continued to deteriorate, even as users were logging over one million hours of game play every day. In March 2004, tensions came to a head. Accusing MOVES of financial improprieties, Wardynski and OEMA ripped the game out of the school and moved it to two new army-controlled locations.

Immediately after taking full control of the game, OEMA signed a licensing deal with the prominent game company Ubisoft to release commercial versions of *America's Army.* (This is where *America's Army: True Soldiers* comes from.) Shrewdly, OEMA had previously trademarked both the name America's Army and the tagline "The official U.S. Army game: Empower yourself. Defend freedom."

At the same time, Wardynski and OEMA resolved to diversify their product in order to increase the game's chances of long-term survival and gain economies of scale in use and development. Because *America's Army* related solely to recruitment, OEMA had, in effect, only one customer. The key, then, was to expand the client base. OEMA decided to develop a private, government-oriented version of the game that could be used for training, military and otherwise. To do so, Wardynski turned to the Software Engineering Directorate (SED) at Redstone Arsenal in Huntsville, Alabama. The SED, a unit of the army's Aviation and Missile Research Development Engineering Center, had a history of developing sophisticated weapons system trainers.

As the private training version of the game took off, it began to feed the public recruiting one. As Wardynski explains, "Say we didn't need a female soldier for the recruiting game bad enough to make one, but maybe the medical people needed a female soldier, or the Secret Service needed a female body type. We'd make it for them and then bring it into the public game. Our governing principle was, *We own everything when we're done* — in other words, we can use it — *and so it has to be of good enough quality to go back in the recruiting game.*"

For the past several years, the training version of *America's Army* has been as popular as the recruitment one. It is the trainer of choice for dozens of military and government organizations, including U.S. Central Command, the United States Special Operations Command, the Secret Service, the Army Chief of Staff, the Association of the United States Army, the JFK Special Warfare Center and School, the Combating Terrorism Center, and the 7th Army. Customized versions of the game also provide training for such equipment as PackBot and Talon robots, the CROWS weapon system, the Javelin missile system, and the Improved Target Acquisition System. Affordability is a key element here; as defense expert Peter Singer points out, "The game's training module cost just $60,000 to develop, but took training in how to operate robots in war to a whole new level."

Life After *America's Army*

These days, Wardynski has mellowed in his opinion of the MOVES team. "Game developers are a lot like artists," he says. "You don't tell them how to paint the painting. You can come to them with a painting of a waterfall and say, 'We're going to do a waterfall,' but don't tell them what kind of colors to use, how high to make the waterfall, any of that junk. It's a different culture, with different objectives, functions, norms. They want to make the most kick-ass game in the world, or the one that has the best graphics, or the one that's got the most realistic effects. We were just way more hands-on than a lot of producers would be."

America's Army launched Mike Zyda out of the Naval Postgraduate School and into the game industry. He now runs his own research lab, Gamepipe, at the University of Southern California. "The things I'm most proud about regarding *America's Army,*" he says, "are, one, it was the first significant serious game. There were plenty of educational games, but *America's Army* was the first one where people said, 'Wow, you can make a *fun* serious game that's also a hit.' *America's Army* also led to the entire Department of Defense deciding to use game developers instead of contractors to make next-generation training and simulation systems. Plus, it was the most cost-effective thing the army's ever built for recruiting."

Though extremely pleased with the game, Wardynski today has mixed emotions about his experience with the project. A few months before he retired from the army, he and I talked in the sunlit living room of his three-story red-brick house at the end of a quiet tree-lined street behind West Point's Lusk Reservoir. As he looked out over the stone remains of the Colonial-era fort bordering his home, he estimated that 60 percent of his time between 2002 and 2009 was spent fighting with Pentagon officials to keep the game and its related properties alive. He blames a nearly fatal heart condition he developed in 2005 on the stress that running the game engendered. After all, he points out, running the game wasn't even his regular job. He was already working sixty hours a week in his capacity as the director of OEMA; *America's Army* was an unpaid assignment on top of that, one that added fifty more hours to his workweek.

Following his 2009 retirement from the army, Wardynski began a new job as chief financial officer in the Aurora, Colorado, public school system. In fall 2011 he became superintendent of schools in Huntsville, Alabama, the city where the training version of *America's Army* is currently managed. Nine months later, the Alabama Council of PTAs named Wardynski its "Outstanding School Superintendent of the Year."

CHAPTER 5

All but War Is Simulation

FOR SEVERAL DAYS EACH spring, an Orlando hotel packed with tourists bound for Disney World hosts an unusual combination of fit army types in forest-green camouflage uniforms and doughy, ponytailed computer geeks in wrinkled khaki pants. The hotel's salmon-pink exterior, its myriad palm trees, its artificial waterfall spilling into a sizable wraparound pool, provide a curious setting for this annual gathering of the defense world's technological elite. Walk through the glass doors of the hotel's booming convention center, however, and familiar names greet you: Anteon, Boeing, General Dynamics, Lockheed Martin, Raytheon. In conference rooms positioned off the main hall, military officials and defense industry representatives attend presentations and tutorials on the military's most popular games and virtual training devices. In the cavernous exhibitors' hall, defense contractors hawk their latest digital wares. Yet there are no booth babes at this particular conference, no sweeping spotlights or clashes of noise between competing exhibitors' booths. By the standards of many military conferences, this one feels almost small-town. Speak with an éminence grise and the conversation will be interrupted by a constant flow of greetings from friendly passersby. This is hardly Mayberry, however: the attendees—whether from the military, industry, or academia—are leading the Pentagon's shift to game-based training.

Defense GameTech, the conference in question, is the brainchild of PEO STRI, the army's gaming and simulation office. PEO STRI is an acquisition and contracting organization, whose mission is to guide product development to meet army requirements and to manage the deployment of training systems. It spends over $3.5 billion yearly on products and services, for the most part by contracting with commercial companies. In keeping with its commercial focus, PEO STRI's current motto, "Putting the power of simulation into the hands of the Warfighter," has the ring of corporate branding. Its previous motto, "All but war is simulation," captures the organization's military-corporate role. Given its location in central Florida's expanding high-tech corridor, PEO STRI benefits from close contact with neighboring military contractors, not to mention such entertainment hot spots as Walt Disney World, Epcot, and Universal Studios. Virtual training and simulation offices for the navy, air force, and Marines are also located nearby.

The stated purpose of Defense GameTech is to advocate the use of video games and video game technology within the Pentagon and to give DoD personnel hands-on learning in game-based training. When I attended a recent GameTech, the panels I observed and the officials I interviewed reinforced Ralph Chatham's and Dan Kaufman's earlier claims that the military had turned to gaming for help in dealing with the new, unexpected roles soldiers have to face. Wars always happen when you least want them to, a former Pentagon official told me, and then militaries spend most of those wars playing catch-up. In the GameTech exhibitors' hall, most contractors were featuring new and improved variations of Ralph Chatham's Tactical Language and Culture Training Systems program: games for teaching soldiers how to interact with local populations. Colonel Franklin Espaillat, PEO STRI's project manager for combined arms tactical trainers, described the change for me: "During the Cold War, the guys that were getting trained on how to interface with the local culture were the senior leaders, because it was part of the Cold War lockstep thing. But now the private, the corporal, who's at the tip of the operation spear, is the one who might be engaging in that local province or with that local tribe. So we have to give him the right cultural training, the right civil affairs training, the right tactical training, which we didn't necessarily do before. We've had to expand training to the lowest possible level. Gaming is what *allows* us to expand that training."

Currently PEO STRI's acquisitions program, Games for Training, centers on *Virtual Battlespace 2,* or *VBS2.* Like the action sections of *America's Army, VBS2* is battlefield simulation, similar in look and feel to a commercial first-person shooter. *VBS2* is an army "program of record," meaning that it will be maintained by the army for as many years as possible before being replaced. (The Marine Corps is also using *VBS2,* primarily as a fire support team trainer.) While army programs of record usually last for twenty to twenty-five years, Games for Training programs last no longer than five years, owing to the speed of innovation in the video game industry. In fact, PEO STRI is currently soliciting bids for *VBS3.* "We're continually inserting things in the current gaming engine to refresh it and make it current with technology," Colonel Espaillat says, "but because gaming in general is a multibillion-dollar industry, we're really just scratching the surface of the overall capability, and trying to leverage the *little* bit of dollars that we get from the army to get a *lot* of capability." In keeping with this strategy, PEO STRI is also using *VBS2*—for which the army paid close to $10 million—for other initiatives. For example, the organization is launching a new product line called *Dismounted Soldier,* which is an attempt to build a 360-degree virtual trainer. Intelligent Designs, the company developing *Dismounted Soldier,* is relying on the *VBS2* gaming engine to power the trainer.

At Defense GameTech, *VBS2* was the Kool-Aid that everybody drank. In presentations, panels, and interviews, military officials and defense contractors alike sang the program's praises. Given that it is now the military's most widely used video game — fully 50 percent of the army is training with it, as is the Marine Corps — the particulars of *VBS2* are worth examining. It is designed to plan mission training and mission rehearsal and to provide education — hence the inclusion of language training, cultural training, and medical training, along with a captains' career course focused on leadership. While the game's graphics aren't quite at the level of today's commercial blockbusters, they are close enough for users to be easily drawn into the action. In *VBS2*, players operate avatars on the ground, in vehicles, in the air, or at sea as they run through scenarios they will encounter in battle.

Administrators (higher-level military personnel) create the scenarios beforehand using a vast library of characters and objects; they also watch and intervene during training from a Tactical Operations Center, which enables them to make real-time adjustments. One of the game's more notable advances is its inclusion of geospecific databases. A soldier training stateside can thus be provided with a database of the exact location to which he will be deployed overseas — the buildings, the roads, the surrounding environment — well before he even arrives. The game also allows commanders and soldiers in war zones to input scenarios from the battlefield (an IED attack, say, or an ambush) immediately after they occur. Within ninety-six hours, those scenarios will be available to the soldiers back home in the States, keeping soldiers who have yet to deploy up-to-date on the opposing forces' latest tactics. In addition, PEO STRI is enlisting soldiers who have returned from Afghanistan and Iraq to develop scenarios based on their battlefield experiences.

VBS2 records players' actions so that administrators can conduct review sessions, known in the military as "after-action reviews" (AARs). This enables leaders to take the soldiers back through the scenarios and point out what they did correctly and incorrectly. The soldiers will then run through new scenarios, in which they can sharpen the skills that need improvement. "Simulation isn't training unless you do a very thorough AAR," Colonel Espaillat says. "That's what makes it training — the feedback you get about the things you did right and wrong." *VBS2* also records audio of the soldiers communicating with each other during the game, which Espaillat says is especially useful when soldiers attempt to deny their errors: "'That wasn't me,' they'll say, but the audio lets you show that it *was* them."

One of the game's distinctive properties is the flexibility it gives administrators: it offers both offline mission editing, which lets them build scenarios before training, and real-time mission editing, which lets them intervene during a training session. Administrators can also input terrain from a variety of shared data files, so they can create virtual worlds that mirror any number of physical locations. (To demonstrate how quickly this could be done, the game's maker, Bohemia Interactive, unveiled an explorable 3-D simulation of Osama bin Laden's Abbottabad complex just days after the raid that killed him.)

At this point, the *VBS2* content library features more than four hundred military and civilian vehicles; hundreds of characters representing at least five national militaries, press agents, and civilians; dozens of weapons; and countless varieties of animals, signs, buildings, natural objects, and paraphernalia, such as alarm clocks and soda cans. The newest iteration also shows variations in avatars' texture and body heat, tides that conform to a region's latitude and longitude, an accurate star field, a face texture editor, and lighting effects that simulate night vision and light blindness. The game's field of terrain can cover as much as 124 square miles and accommodate as many as 256 players at once. In addition, its open platform — which, among other things, allows soldiers in the field to plug in the latest battlefield data, just as Ralph Chatham's DARWARS did — gives the game a wide degree of flexibility. Since 2010, user input has grown by more than 1,200 percent.

A U.S. Army Research Institute study of 165 soldiers found that "the training [positively] impacted how well the Soldiers felt they could work together as a team, as well as their attraction to the unit, including their attraction to their tasks and other group members." The soldiers in the study felt better prepared for tactical convoy operations following their training.

Beyond the ARI study, the military has little hard data on whether the game produces physiological and emotional responses similar to those in battle itself. "We haven't done any formal, satisfactory analysis," Leslie Dubow, PEO STRI's project director for Games for Training, told me, "just because it's so difficult to do. So we don't have hard data; we have anecdotal data about how effective *VBS2* is." Marine platoon commander First Lieutenant Roy Fish reports, for example, "You can't simulate the dust, dirt, heat and stresses that you inevitably feel in combat situations, but I think [*VBS2*] gets as close as you're ever going to get to Afghanistan." Fish first encountered *VBS2* in 2008, when he was training at Camp Lejeune, North Carolina, and he insists that the game has saved his soldiers' lives in Afghanistan. "Every time we go outside the wire and react to an I.E.D. or small-arms fire, it all translates to what we did [in *VBS2*]," he says. When his troops finish a round of *VBS2* training, he continues, they are "sweating from head to toe. It's amazing how realistic it was. It's literally the same terrain."

More Than a Game

To find out how *VBS2* training works in practice, I paid a visit to Fort Campbell, Kentucky, home of the 101st Airborne Division Screaming Eagles — and, since early 2011, site of the Kinnard Battle Command Center's cutting-edge Virtual Training Facility (VTF). Located on Air Assault Street, the VTF's plain brick exterior gives little hint of the technological focus inside. Enter the building, however, and you find a windowless classroom containing rows of tables lined with computers — enough to fit a company-sized group of one hundred soldiers at a time. Across the hallway sits the Tactical Operations Center, smaller than the classroom but similarly filled with computers, a digital projection system, and a sound system. From here commanders and administrators, unseen by the trainees, manage the training occurring in the classroom, adding a *Truman Show*–like element to the proceedings. Built for the meager sum (in military terms) of $750,000, the

VTF is where soldiers from the 101st spend up to a week training on *VBS2* before deploying to Afghanistan.

At the VTF I met with virtual team chief Adam Williams — an ex-Marine, Williams works for the defense contractor SAIC — and Jeff "Beast" Jackson, a gregarious, heavily muscled former drill sergeant with a quick wit and a deep, occasionally impenetrable southern accent. Both men exhibit a combination of seriousness and playfulness that befits their roles as military trainers who work with video games.

Depending on the commander's wishes, some units train on *VBS2* until the day they deploy overseas, while others do it earlier in the training cycle. (Some units even take *VBS2* along on their deployments.) Before a given session at the VTF, the company's commanders meet in the Tactical Operations Center with Williams and his technicians. The leaders tell Williams exactly what they want the day's training to focus on and the specific location in Afghanistan or Iraq where it should be set. Most commanders want to focus on conducting convoy and dismounted operations, including small-arms fire, experiencing near and far ambushes, requesting casualty and medical evacuations, reacting to IED attacks, and requesting bomb-disposal teams. Some units pay Afghan interpreters to join the training sessions as role players. By the time Jackson has finished describing how to use the keyboard, Williams and his team have pulled up the proper scenario (or scenarios) on *VBS2*.

If a scenario or setup is particularly innovative, Williams also posts it on the Army Training Network, an online tool for army instructors. "The best part," he says, "is that you get to network out. Hey, what's Fort Lewis doing? What's Fort Drum doing? Each site is focusing on different things, so that really helps us grow. I'll ask Fort Bragg, how are you using *VBS2*? What's worked for you? Do you have any scenarios that I can share?"

When a company of soldiers sit down at their computers for their training session, Jackson starts with a lesson in keyboard functionality, showing them how to operate their avatars. "A lot of times, soldiers don't understand how to put their fingers on the keyboard and type and things like that," he explains. "So you say, 'Hey, if you want to walk forward, you have to press and hold the *W*. If you want to do a basic run, you tap the *W* twice and hold. If you want to run fast, push down with the left ring finger on the left hand, which holds the shift key, and tap the *W* twice. You can make the avatar slide to the left by pressing the alpha [*A*] key, or straight to the left by hitting the delta key. You make him get in the crawl position by hitting the Zulu key. There's also crawl forward, crawl backward, land down, come up, roll left, then right. When you're looking at the soldiers, you can see some of their expressions, the ones that aren't keyboard-savvy, they don't like it. So you have to make sure you go slowly."

Once a training session begins in earnest, soldiers must navigate the various scenarios in the game while sticking to the standard operating procedures and battle drills they've already learned. And here, predictably, is where the trouble begins. The hardcore gamers among the soldiers invariably begin their sessions by trying to operate outside the system. They make their avatars run around shooting everything and everyone in sight, including their commanding officers. This is where Jackson steps in. He begins gently enough. "Normally, if I see gamers not doing exactly what they're supposed to do, I'll talk to them, or I'll get the leader to talk to them, and explain that they're not supposed to play around. I'll put my hand on their shoulder and emphasize that what they're doing is *training*, that it's something they can *learn* from. Because they have to go over that pond to Afghanistan. Here I can push restart and begin the scenario again and they'll still be alive. But I say, 'When you cross that pond, there is no restart, because those bullets coming at you are real.'" If the soldier continues to play around, Jackson simply relies on his intimidating presence to adjust that soldier's attitude, as he did back in his drill sergeant days. "He can just draw up and give the soldier a look," Williams testifies, "and the person immediately gets in line."

Williams can also intervene in the game from his position in the Tactical Operations Center. "If I see them doing something that I really don't like — say we're in a training exercise and they're shooting when they're not supposed to be shooting—I have the power to just go in the game and strip them of their ammo, kill them, and then put them in a forest twenty klicks away. They get the idea pretty quickly that way."

Scott Rosenberger, another VTF trainer, estimates that 80 percent of the soldiers who go through the facility play video games regularly. (A much smaller group, he says, are hardcore gamers.) After the initial horseplay, a sense of realism usually descends on the training sessions, in part because of *VBS2*'s replication of real-world dynamics. "If you had to run from here across the street and run back, you'd be breathing hard," Jackson says. "And if you then had to take your weapon and aim it at that wall, you'd be breathing even harder. And that's just how it is for the avatars in *VBS2*. To control your breathing, you have to right-click and hold down. But you can only hold it for so long. In real life, if you aim your weapon, in about ten seconds your eyes will start getting blurry and then you'll have to take your cheek away from it, take a breath, and start over. Same thing in *VBS2*. The avatar will start trembling and he will lose sight. So you have to recock and do it again. Reset, take your hand off the right mouse button, and then reengage the target."

With *VBS2*, each soldier can switch between first-, second-, and third-person mode to view the action. Headphones transmit the ambient noise of the battlefield, including radio transmissions, the sound of helicopter rotors, gunfire, and explosions. Because all one hundred soldiers are linked together in the simulation, the same hierarchies that would exist on the battlefield soon emerge in the classroom. Soldiers who are better at moving their avatars, for example, help those soldiers whose avatars fall behind, while soldiers who have a better grasp of standard operating procedures help soldiers who are still learning them. "You really find out who the leaders are within a group, regardless of the troop commander or the vehicle commander sitting there," Jackson says. "You end up having your gunners or one of your passengers step up to the plate and tell the troop commander what to do, because the TC may be nervous, and he or she can't handle all that pressure." The platoon leaders, who observe the training from the Tactical Operations Center, often force the issue. "The leader may tell us to kill the convoy commander, just to see who steps in and fills that vacuum."

Williams cites this as perhaps *VBS2*'s greatest benefit. "It allows commanders to evaluate their junior leaders, to see their strengths, see who is hesitant, who is aggressive. *Okay, I just taught you this directive. How well can you execute it?* Or, *Are my NCOs stepping up, or are they getting walked all over? As a commander, is my intent being executed right now?* And we've seen that when you have leadership at that high level driving things, it's even more effective. Because when you have a full-bird colonel sitting in a room like this one, watching a platoon leader operate his platoon on the ground, that platoon takes him pretty damn seriously. Because they're not learning anymore; now they're executing." *VBS2* also gives commanders greater freedom than traditional muddy-boots training can. "The commander might say, 'Hmmm, I want to do an air assault from this location, dismount here'—you can start trying things that may be too dangerous to try out there in the field, because of the amount of fires that are involved or the coordination with aircraft. That's when they start doing it here, virtually," Williams says. "It's great if you have real pilots playing their roles, because that gives that young platoon leader, those platoon sergeants, those NCOs, an opportunity they very rarely get to talk face-to-face during the after-action review with a pilot. And the pilot can say, 'Hey, when you're in theater, this is how I want you to talk to me, this is the information I need.'"

As a training session wears on, Williams continues to guide the action—in response to the commanding officer's wishes—from his perch in the Tactical Operations Center. "Say the players are just not generating the type of response their commander wants to see, and so he decides to ping them with mortars. I can literally just hit Artillery Strike, point where I want it to go and what kind of munitions I want to use, how many guns, how many rounds, the impact radius, and then say go. And the players will react: 'Oh my god! Incoming!' Or let's say they're flying in an Apache. I can take that Apache, crash it, and generate a *Black Hawk Down* scenario."

Williams then proceeds to demonstrate exactly that—except that the helicopter fails to explode on impact as it's supposed to. "One thing about DARWARS that I liked more than *VBS2*," Williams admits, "is that it was more stable. There are still glitches in *VBS2* where for some unknown reason the system will crash. Software problems, stuff that I can't touch."

Even with the occasional software glitch, *VBS2* manages to generate a level of emotional intensity that mirrors real-world conditions. "I've seen people cry," Williams tells me. "I've seen certain people in certain positions get so stressed-out and upset—and we're purposely stressing them out—that they have stormed out of here and sat in their car for fifteen minutes, because they just couldn't take it. I've seen people who were so upset with their performance . . ." He stops. "You get all sorts of reactions."

Williams says things can get especially heated during the after-action reviews. "We don't lead the AAR. The NCO or the officer, whoever's in here, leads the AAR. And the soldiers will argue with each other. 'Well, shit, I told you to frickin' take your squad and go up that hill!' 'Bullshit!' You let them handle that. It may sound aggressive, but that's how they talk to each other. If that helps them figure things out for later, then it's a success in my book. They don't necessarily have to like what happened in here."

This displeasure on the part of soldiers can extend to the training session as a whole. "Sometimes you get this poor E3 whose only job in the frickin' scenario was to get blown up, and he's pissed! He's like, 'This sucks! I sat here staring at a black screen for three hours!' That can be construed as looking negatively on us, but in reality it's what the unit wanted. So I always try to focus on what the leadership wants to do. I always ask the commanders, 'Who's your target audience? Is it the whole platoon? Do you want to evaluate your squad leaders? What do you want to see happen here?' And we always get positive remarks on that."

With other military video games in mind, I ask whether any of the training focuses on instilling army values.

"Yes, if the intent is to reinforce things like escalation of force and rules of engagement, rules of land warfare, which we always do," Williams says. "You know, 'Okay, you just frickin' mowed down that person's cow, we need to talk about that. It's ha-ha funny in the game, but you just took away that person's livelihood. How much cash did that cost your battalion commander, who had to pay off that village elder?' We talk about second- and third-order effects. 'This car was just trying to drive around the convoy — why did you feel the need to drop eighteen rounds of fifty-cal through their windshield? Is that the rules of engagement?' You talk them through it, and they leave thinking, *Okay, what did I learn? I learned that maybe shooting Ahmed's cow is not the best move of my career.*"

"A lot of soldiers enter the training sessions thinking they're gonna be fun, just a chance to play around on video games," Jackson explains. "And I've seen how shocked they are when they're killed in *VBS2*, when they can't just reboot. I've seen them start to make connections between what they're doing in the simulation and what they'll be doing in Afghanistan. For a lot of these kids, playing *VBS2* is the first time that they realize they might die."

CHAPTER 6

WILL Interactive and the Military's Serious Games

THE GAME GOES LIKE THIS: You are Specialist Kyle Norton, a nineteen-year-old midwesterner whose life has begun to spiral downward following a tour as a bomb-disposal technician in Iraq. Already beset by financial difficulties, you receive a surprise e-mail from your fiancée, who announces that she has become pregnant by another man. Still reeling from this news, you learn that your best friend has just been killed in an ambush. As these scenarios unfold, questions appear on your video screen, prompting you to decide whether, as Norton, you should seek help for these issues. Depending on your responses, Norton either becomes suicidal or begins to heal.

This game, *Beyond the Front,* is the brainchild of WILL Interactive, a Maryland-based serious-games company that has the military as its major client, accounting for about 75 percent of its business. Founded in 1994 by former secondary-school teacher Sharon Sloane, WILL differs from its competitors in that it utilizes what it calls Virtual Experience Immersive Learning Simulations, or VEILS — live-action, interactive movies in which users become the lead characters. The games are based on real events, use real actors, are shot on location, and last between two and three hours. Similar to the old Choose Your Own Adventure book series, each game offers players more than eighty "moments of decision," every one of which affects the story line and the outcome (each story contains at least a thousand possible permutations). For her work, Sloane received the 2009 Women in Film and Video's Women of Vision Award, along with the U.S. Army's 2008 and 2009 Distance Learning Maverick Awards.

Like *VBS2* and *America's Army,* WILL's products are examples of what game aficionados and scholars call "serious games"; the phrase refers to any game designed for something other than entertainment. Though it may be entertaining, a serious game's primary function is to educate the player or help him or her solve a problem. Serious games are used extensively in education, health care, city planning, science, engineering, and emergency management, but their primary maker and user has always been the military.

The first serious games were in fact war games. Many scholars point to Kriegsspiel, a nineteenth-century Prussian military officer training game, as a direct antecedent of today's serious games. The term itself seems to have first appeared in Clark Abt's 1970 book *Serious Games,* in which he declared, "These games have an explicit and carefully thought-out educational purpose and are not intended to be played primarily for amusement." Though Abt was referring to card games and board games, his definition remains relevant in the digital age, when almost all serious games are simulation-based computer and video games. (The army's never-used 1980 game *Battlezone* is generally considered the first contemporary serious game.) In 2005, following his experience with *America's Army,* Mike Zyda updated Abt's definition by defining a serious game as "a mental contest, played with a computer in accordance with specific rules, that uses entertainment to further government or corporate training, education, health, public policy, and strategic communication objectives."

In the field of education, paper-based serious games were common in the 1960s and 1970s, before the Back to Basics movement relegated most of them to the dustbin. With the widespread introduction of computers into the public school system in the 1980s, computer-based serious games such as *Oregon Trail* and *Math Blasters* became popular in classrooms nationwide, followed in the 1990s by more advanced games such as *Dr. Brain,* which teachers often passed over in favor of the newly arrived Internet.

Outside the public school realm, the most prominent serious-games movement originated at the Woodrow Wilson International Center for Scholars in 2002. Headed by Ben Sawyer and David Rejecsk, the Washington, D.C.–based Serious Games Initiative promotes the development of digital games in the realms of policy and management and has led to such offshoots as Games for Change, which focuses on social issues and has clients such as USAID, the World Bank Institute, the American Museum of Natural History, and CURE International; another offshoot, Games for Health, is sponsored by the Robert Wood Johnson Foundation and focuses on improving health care. But Sawyer acknowledges that the military was and is by far the most prominent player in the movement, in regard to both early adoption and continuing funding, development, and use.

And WILL is the most prominent creator of serious games for the military, particularly games that address the psychological and emotional issues service members might face. The games that WILL develops for the military are primarily non-combat-related, focusing instead on what Sloane calls "high-stress situations, high-risk situations, and emotional issues," including suicide prevention, mental health, sexual assault, off-duty behavior, ethical decision-making, and reintegration into civilian life. The ability to address such a broad range of noncombat issues is part of why WILL has become the go-to serious-games designer for the military.

The Emotional Component

A graduate of Boston University, WILL's Sharon Sloane began her career as a high school English teacher before moving on to stints as a counselor at both the high school and the community college level. She soon found, however, that her real interest lay in instructional design—specifically, in developing training materials. After leaving the classroom, she worked as an independent consultant doing product development; her primary materials were videotapes and an early form of computer-based training.

In the late 1980s, Sloane consulted on a project that involved developing interactive video disks—the disks, a new technology at the time, resembled large phonograph records—to train emergency-room personnel. In the series Sloane worked on, trainees played the roles of doctors or nurses treating patients suffering from heart attacks, seizures, or gunshot wounds. With each case unfolding in real time, trainees had to make diagnoses, order medications, and perform other urgent tasks. Despite the cumbersome, expensive technology, Sloane found the experience of working on the simulations—of immersing doctors and nurses in life-and-death situations—"profound." The technology itself may have been primitive, but to Sloane, the learning method behind it was potentially groundbreaking. "It really hadn't been done before," she says, "in terms of making learning a real-time, first-person experience."

For Sloane, the experience was a wake-up call. At the time, the field of modeling and simulation operated on the assumption that if you gave people information, that information would automatically translate into behavior and skills. Sloane realized that this assumption was only partly true. Information wasn't enough to change people's behavior; true learning required an *emotional* component.

After founding WILL in 1994 with Lyn McCall, a former army colonel with experience in modeling and simulation, and Jeffrey Hall, a screenwriter, Sloane developed a new kind of digital game—the aforementioned VEILS, which were based on slice-of-life true stories. Sloane had asked herself, how can we use video and interactivity to engage people emotionally and cognitively in order to change their behavior? Her solution was to create games that were in essence interactive movies; their first-person perspectives allowed players to "be-

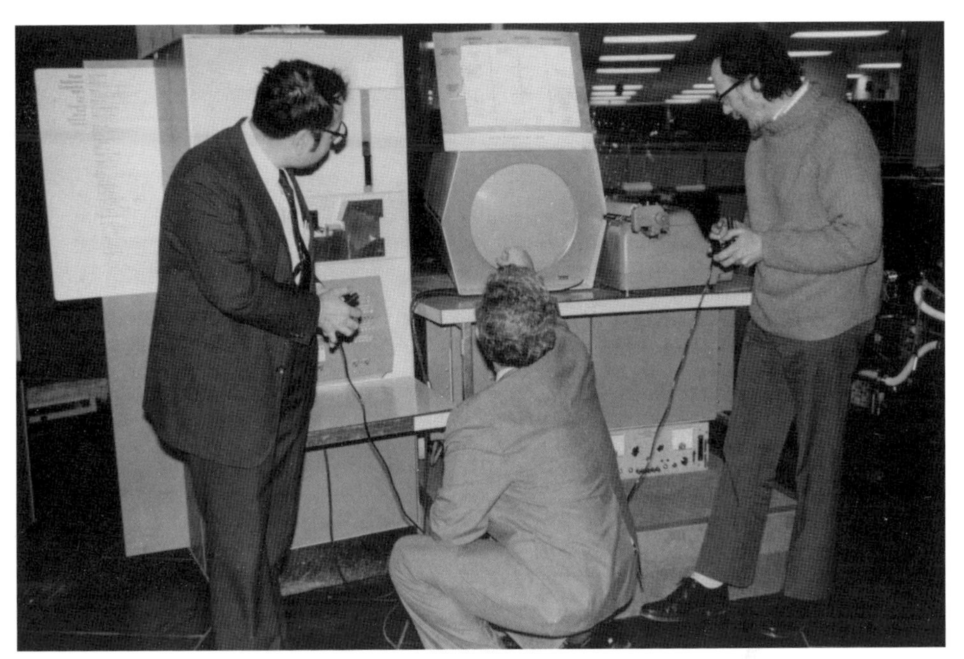

Spacewar! being played on a PDP-1 at the Massachusetts Institute of Technology. Arguably the first digital game, *Spacewar!* was invented in 1962 by twenty-three-year-old math major Steve Russell and his cohorts in the fictitious Hingham Institute Study Group on Space Warfare, a collection of like-minded, Pentagon-funded engineering graduate students at MIT.

Courtesy of the Computer History Museum

Part of the IBM AN/FSQ-7 computer that was the backbone of the U.S. Air Force's Semi-Automatic Ground Environment (SAGE). The SAGE project led to significant advances in core memory, keyboard input, graphic displays, and digital communication over telephone lines.

Scott Beale / Laughing Squid

DARPA (Defense Advanced Research Projects Agency) headquarters, Alexandria, Virginia. DARPA is the Pentagon's technology research lab; along with funding the creation of numerous weapons systems, DARPA is responsible for the invention of computer networking, as well as ARPANET, an early version of the Internet. It has also been a primary sponsor of research in computer science and artificial intelligence over the past several decades.
Arlington News / ARLnow.com

FlatWorld, shown here at the Institute for Creative Technologies. One of the ICT's earliest projects, *FlatWorld* was conceived of and designed by a combination of video game designers, special effects artists, research scientists, and Pentagon personnel working together to create the army version of *Star Trek*'s fictional "holodeck," a virtual space that can be programmed to mimic a wide array of three-dimensional settings. *USC Institute for Creative Technologies*

Colonel Casey Wardynski at West Point. For more than a decade, Wardynski ran the U.S. Army's Office of Economic and Manpower Analysis; while in this position, he created *America's Army,* the world's first-ever military-developed video game. He is now superintendent of schools in Huntsville, Alabama. *Courtesy of the author*

A screenshot from *America's Army*. From 2002 to 2008, *America's Army* was one of the top ten online games in the world. Originally designed as a recruiting tool, the game has since been repurposed as a training tool for the military and numerous government agencies.
U.S. Army Office of Economic and Manpower Analysis

Tactical Iraqi, released in 2005, trained soldiers in Baghdad Arabic and Iraqi culture. The game's emphasis on training even junior personnel in intercultural difference represented a new phase for the military; in the past, understanding the culture in which one was fighting would have been seen as the purview of the State Department, not the armed forces.

© Alelo Inc. Used by permission

A soldier training on *VBS2.* Currently the military's most widely used video game, *VBS2* is designed to plan mission training and mission rehearsal, and also to provide education—hence the inclusion of language training, cultural training, and medical training in the hardware, along with a captains' career course focused on leadership. *Catrina Francis (USAG Fort Knox)*

A screenshot from *VBS2*. *Bohemia Interactive Simulations*

Filming a scene for a WILL Interactive learning game. Founded in 1994 by former secondary school teacher Sharon Sloane, WILL differs from its competitors in that it utilizes what it calls Virtual Experience Immersive Learning Simulations, or VEILS—live-action, interactive movies in which users become the lead characters. The games are based on real events, use real actors, are shot on location, and last between two and three hours. Similar to the old Choose Your Own Adventure book series, each game offers users more than eighty "moments of decision," with every decision affecting the story line and the outcome. *Courtesy of WILL Interactive*

Matthew Pennington at home in Maine. After being injured in battle, Pennington found that playing a WILL Interactive learning game helped guide him toward treatment for his PTSD.

Todd Heisler / New York Times / Redux

Skip Rizzo in his office at the Institute for Creative Technologies. Rizzo is one of a small but increasingly influential group of psychology researchers working in the discipline of virtual-reality exposure therapy (VRET). Since its origins in the early 1990s, VRET has proven remarkably effective at treating anxiety disorders; initial studies indicate a "cure" rate of between 70% and 90%.

Courtesy of the author

Soldiers in Iraq demonstrating Virtual Iraq. The brainchild of Skip Rizzo, Virtual Iraq—and its more recent counterpart, Virtual Afghanistan—are the most widely used virtual-reality exposure therapy treatment programs in America. Wearing a head-mounted display, VRET patients are placed in immersive, interactive environments designed to represent their traumatic memories, ensuring that they can confront their experiences without having to conjure up the memories themselves. *Courtesy of Greg Reger*

SimCoach's Bill Ford. In *SimCoach,* virtual humans act as advisers and sounding boards for members of the military community who may be suffering from such issues as depression, stress, substance abuse, suicidal ideation, brain injuries, and relationship difficulties. One of the primary goals of the application is to break down the barriers to care that have traditionally dogged the military—most prominently, the belief among soldiers that they will be judged negatively if they admit to needing help. *USC Institute for Creative Technologies*

Students at New York City's Quest2Learn, the nation's first public school to feature a curriculum based entirely on the principles of good game design. According to the school's mission statement, "Games work as rule-based learning systems, creating worlds in which players actively participate, use strategic thinking to make choices, solve complex problems, seek content knowledge, receive constant feedback, and consider the point of view of others."

Photo by Gillian Laub

U.S. Cyber Command, which is charged with centralizing and coordinating the various cyberspace resources that exist throughout the military. Headed by National Security Agency director General Keith Alexander, Cyber Command is responsible both for protecting the Defense Department's information infrastructure and for developing new offensive and defensive cyberwar capacities across the services as a whole. *U.S. Air Force / AGE / F1online.de*

The Chinese army's *Glorious Mission*. Reportedly modeled on *America's Army*, the game requires players to complete basic training before advancing to online team combat. After players complete training and combat, they move into a third stage, which re-creates what a Chinese news report calls "the fiery political atmosphere of camp life." *Giant Interactive Group*

come" the lead characters and make decisions that would affect the stories' final outcomes. This interactivity was key to making the players feel an emotional investment that would lead to actual behavior modification.

In 1998, WILL's first contract from the military resulted in *Saving Sergeant Pabletti,* a game that focuses primarily on sexual harassment. The army had approached WILL in the wake of the Tailhook scandal, in which dozens of navy and Marine Corps aviation officers were charged with sexual assault and improper conduct following an annual symposium in Las Vegas. In *Saving Sergeant Pabletti,* a team of twelve trainees in the woods of South Carolina is suddenly rendered leaderless when Drill Sergeant Pabletti is struck by a hunter's stray bullet. The trainees need to make a stretcher to carry Pabletti to safety, but sexual and racial prejudices prevent them from working as a team, and Pabletti eventually dies. In playing the game, users are offered the choice of six possible characters. With each character, they are taken back in time to the week before the incident. They are then required to make "values-based decisions" for their fellow soldiers in order to change the way the team as a whole thinks and acts. In so doing, WILL claims, players are addressing not only sexual harassment but also issues relating to leadership, ethics, and racism.

The initial response from soldiers to *Saving Sergeant Pabletti* was positive enough that years later, following the 2004 Abu Ghraib prison scandal, the army ordered incoming soldiers to play the game on their flights to Iraq. In both cases, the military was using the game partly for public relations purposes. As Sloane says, "The army wanted to show the public they were taking care of the sexual harassment issue."

Pabletti led to several more army contracts for WILL as well as to inquiries from the other services. In 2006, as the military reported its highest suicide rate in decades, WILL received its biggest contract yet: to build a game that would combat this alarming uptick in soldier deaths and help to remove the stigma that many soldiers associated with seeking mental health care. The result was *Beyond the Front,* the game featuring Specialist Kyle Norton. *Beyond the Front*'s unique vantage point — players actually take on the identity of someone who is suicidal — made the game a standout in its military context. The army was so pleased with the results that in 2009 it made *Beyond the Front* mandatory for all active-duty, National Guard, and reserve units. Every soldier in the army played the game, and the feedback WILL received was again uniformly positive. Soldiers appreciated the game's rawness, the bluntly realistic manner in which it depicted the stresses, even the horrors, of postcombat life.

With its profile within the military now higher than ever before, WILL received a contract to develop *The War Inside,* in which players take on the roles of four different characters: a young soldier with post-traumatic stress disorder; a sergeant who can no longer connect with his wife and child; a spouse trying to understand her partner's emotional difficulties; and a platoon sergeant working to create an environment that encourages his soldiers to seek treatment for mental health issues. In a related game, *The Mission to Heal,* players can also be a social worker experiencing compassion fatigue, a case management nurse teetering on the edge of burnout, and a squad leader trying to overcome his own prejudices against mental health care.

Once again the military was happy with the results. In the meantime, however, a new issue had arisen. "What about National Guard and reservists?" the Pentagon asked Sloane. "They lack support, they're not on a base, they come back home and they've lost their jobs, their suicide rates are shooting upward." In response, WILL developed *The Home Front,* which emphasizes suicide risk awareness and suicide prevention among guardsmen and reservists.

Currently WILL is focusing primarily on what the army calls "Comprehensive Soldier Fitness." ("Strong Minds, Strong Bodies," goes the cheesy official tagline.) The idea is that building resilience in soldiers before they deploy can lessen the impact of combat stress, including, potentially, post-traumatic stress disorder. In outlining the program, the army identified five separate areas of importance: family, and physical, spiritual, emotional, and social well-being. WILL has now developed a series of sixteen games covering the first two areas, family and physical well-being.

In the first game, *Single Parenting,* players take on the role of Specialist Bridget Franklin, a young soldier who has just had a baby. When Franklin receives orders to deploy, she must figure out how to balance her professional obligations and career requirements with her parenting obligations. With no partner to help out at home with the baby, what is her family care plan? Are her parents a potential resource? (Maybe, but they have serious health issues.) Will the baby's father play a role while she's deployed? Should she leave the army instead of leaving her baby? These are among the questions soldiers must resolve as they play the game.

In a second game, *Blended Families,* players take on the role of Keith Earhardt, a staff sergeant with a daughter from a previous marriage, another daughter from his wife's previous marriage, and a son from his current marriage. At stake in the game are two primary issues: how the families will blend together and how they will fare during Earhardt's deployment.

WILL is also focusing on military ethics. In one new game, players, as unit commanders, must figure out how to reconcile their orders with their personal value systems. In each scenario, they have to balance loyalty to the other people in their unit, loyalty to the army's mission, and loyalty to their own morals. The scenarios are nuanced. In one, the player's unit comes upon a wounded civilian; the local culture is such that the unit shouldn't intervene, and yet if the soldiers don't provide medical assistance, the person will probably die. If they do intervene, players must also determine whether to use military resources, which are supposed to be for soldiers only, to help the civilian.

Sloane says that the continuing evolution of WILL's games for the military matches the issues that are "keeping the leadership up at night." Ten years ago, she says, "we wouldn't have done PTSD and suicide prevention. We wouldn't have done training for warrior transition units or soldiers who are severely injured physically." At the moment, the leadership's primary concern, Sloane says, is "what happens when soldiers return from their deployments — the strains and the stresses on the family members." As we'll see, more and more soldiers are fighting these "invisible wars" at home.

A Soldier's Story

When Matthew Pennington enlisted in the army at age seventeen, he was escaping an alcoholic father and a future of dead-end jobs. A child of divorce, he had grown up dividing his time between rural Maine and Fort Worth, Texas. The stability and discipline of the army appealed to him, but more than that, the idea of regularly preparing for different missions seemed to promise that his life would never grow stagnant. Pennington was easily bored, and the army seemed like a place where his desire for continual change would be rewarded.

In 2002 he deployed to Afghanistan, where, as a Signal Corps operator, he helped build a communications network from the allied air base at Bagram. The tour was engaging and without incident. Upon his return to Fort Bragg, Pennington quickly volunteered to go back overseas, this time to Iraq with the 2nd Armored Calvary Regiment. Though a friend of his was killed during this tour, Pennington still enjoyed his time in a war zone. When he returned to Fort Bragg, he volunteered to head back to the fight again. He received training as a machine gunner and then redeployed to Iraq, where his unit provided logistical support for resupply convoys.

By now it was late 2005. Almost as soon as Pennington touched down in Iraq, he could tell that the war's tempo had changed. Previously the violence had seemed sectioned off, reserved for certain locations and towns. This time it was clear that the violence could — and would — strike anywhere at any time. His superiors informed him that members of his unit had only a 65 percent chance of surviving. "It wasn't a matter of *if* [the violence] would get you," he says, "it was *when.*" Yet as a self-described "adrenaline junkie," Pennington thrived on how "pumped up" the danger made him feel.

One night, however, Pennington had a bad feeling. "It was one of those missions where everything seemed to go wrong that could go wrong," he explains. His convoy that night included a number of fuel and ammunition trucks, vehicles that had become frequent insurgent targets. Pennington was in the driver's seat of the lead Humvee, a position he disliked; he preferred being on top of the vehicle, behind the machine gun. As they traveled by night from Balad to Tikrit, the convoy was plagued by fog. While motoring through the insurgent stronghold of Samarra, Pennington's vehicle abruptly lost its lights. With only one low beam remaining, Pennington arrived at a 90-degree turn in the road; off to the left, several insurgents opened fire. As he steered into the turn, he saw an IED in the road directly ahead. With no time to avoid the blast, he attempted to maneuver his vehicle so the engine, not the cab, would take the bulk of its force. A moment later the IED exploded, sending fire and thick smoke billowing through and around the vehicle.

Pennington tried to slam on the brakes, but there was no response: looking down, he saw that his left foot was gone, blown off by shrapnel that had pierced the vehicle's floor. His right leg, too, was shredded, and his lungs were badly scorched from the fire and smoke. With no other options, he steered the front of the vehicle into a large dirt mound on the side of the road. As he waited for the convoy's rear vehicle, he placed tourniquets on his legs. He then picked up his gun, trained it on the nearby fields, and began firing at the insurgents closing in.

Pennington spent the next year in a haze of medication and rehabilitation at Walter Reed Army Hospital in Washington, D.C. His only concern was his physical recovery: he wanted out of his wheelchair. He had been diagnosed with post-traumatic stress disorder, but his doctors had placed him on such a strong cocktail of opiates that he had little sense of any emotional distress. When he was released, he and his wife returned to central Maine, where Pennington, with his doctor's permission, stopped taking his medications, because he was tired of how groggy and apathetic they made him feel.

In the weeks after moving home with his wife, Pennington fell into a deep depression. He had planned on spending his life in the army, and now that option was gone. He hated his carbon-fiber leg and the cysts it raised on his stump. He hated that whenever he went out in public, people approached to thank him for his service or to share their opinions on the war. "A lot of [soldiers] just want a break" after being at war, he says, "and it's really hard to get one." He began heading to stores later and later in the day; eventually he tried to avoid going out in public at all.

At the same time, his relationships with his wife and friends were rapidly deteriorating. "I was acting like a prick and on real high alert," he recalls. "A lot of people told me this, but I refused to listen." No matter how many times people informed him that he was acting different from his old self, Pennington did not believe them. Self-medicating with alcohol to dull his depression, he seemed on the verge of becoming a full-blown alcoholic. From his perspective, every time he opened his mouth to speak it only made things worse, so he just kept quiet. He had never been one for sharing his feelings anyway; now he simply stopped communicating. With their marriage on the brink of collapse, he and his wife separated three times. One night, drunk and at the end of his rope, Pennington drove his car directly into a brick wall.

This would be the rest of his life, he thought: anger, isolation, alcoholism. But one day, in 2009, a friend forwarded him a casting call from an undergraduate filmmaker at New York University. The filmmaker, Nicholas Brennan, needed someone to play a veteran from Maine who had lost his leg and was suffering from PTSD. Though he had never thought about acting, Pennington was struck by how closely the character's life mirrored his own. As low as his confidence was at that point, he thought acting in a film might "be so out of the normal that it would force me to deal with things." Pennington's desperation overcame his apprehension, and he typed out a moving e-mail to Brennan: "I was injured in 2006. It resulted in a personality change for the far worse."

Pennington wanted to heal, yes, but part of him was also searching for a reason to give up. A conservative, he thought spending time with liberal NYU undergrads was likely to confirm his impression that America wasn't accepting of veterans. Point proved, Pennington could then spend his life sitting angrily alone in his house, convinced of his own righteousness.

Instead, the three days he spent working on the film showed him that even if people were opposed to the conflicts in Iraq and Afghanistan, they were not against the veterans themselves. More important, the process of playing Connor, the film's protagonist, illuminated his own struggle with PTSD. For the first time Pennington began to grasp how serious the issue was.

The next step in Pennington's healing came after the fifteen-minute film, *A Marine's Guide to Fishing*, was completed. Following an initial screening in Portland, Maine, Brennan and Pennington traveled to the 2011 GI Film Festival in Washington, D.C., to present their work. There, on the third night of the festival, in a movie theater inside the U.S. Capitol, Pennington encountered the work of WILL Interactive.

Sharon Sloane began WILL's presentation that night with an overview of the VEILS learning system. She then screened *The War Inside*, the game about veterans' reintegration into civil society following deployment. The audience members that night were given handheld devices so they could play along with the game on the big screen.

As Pennington played through the various characters in the game — an army specialist suffering from PTSD; an officer who no longer relates to his wife and child; a spouse trying to come to terms with her husband's negative behavior — he felt that he was encountering his own life from striking new perspectives. Playing the spouse, for example, he says, "was like watching my wife." The surge of recognition was like a slap in the face: "I thought, 'Wow, okay! This makes so much more sense to me now.' I was able to go back [home] and talk about things more effectively."

A key reason that Pennington responded to WILL's work is that it steers away from what he calls a "direct, confrontational" approach. *The War Inside,* he says, doesn't accost veterans with cries of "Hey, what's wrong with you?" Instead, it enables them to identify with the characters. In doing so, veterans "can turn the light on for themselves. [That approach] makes everything much more effective . . . It's able to open your eyes to see how you may be going about things and why it has proven ineffective for you."

Pennington says that when veterans return home from war, "confrontation is just a quarter-second away — with combat stress, one of the symptoms is that you're on an adrenaline high, and almost anything can trigger it. It could be a car driving past you too fast, and six months ago, a car couldn't come within a hundred and fifty meters of you because it's considered a threat. Little things like that."

Too often, he says, people take an overly forceful approach to treating veterans. This method has it all wrong, he believes: "With veterans who are undergoing withdrawal symptoms, or are maybe even delusional and not seeing their own symptoms, [a confrontational approach is] just going to push them further away." With a digital game-based approach like WILL's, he says, "there's no human-to-human interaction, so [veterans are] able to go through the thought process themselves instead of feeling like they're being told what's wrong with them . . . They're able to come to their own conclusions about it."

Pennington was so impressed by WILL's work that he wanted to spread the word to other veterans. To do so, he enlisted the help of the Coalition to Salute American Heroes, a nonprofit organization where his wife works. The organization holds a monthly training seminar called "Hands of Hope," where a licensed clinical psychologist meets with veterans to talk about their issues. Pennington knew the program director personally; he told her about WILL Interactive and the GI Film Festival and then connected her with Sharon Sloane. Ultimately, "the coalition was able to show *The War Inside* to a lot of its associates and families," Pennington says, "and they're all wounded combat veterans from the current conflict. And they all just ended up falling right in love with it. And they were able to take a lot out of the experience as well, because like me, they were able to see things for themselves" and come to their own conclusions.

"We all have self-awareness," Pennington reflects now. "We all know how we sit with ourselves. And we have our self-identity as well, how we see ourselves. And that's why I [like WILL's games], because through the process, you understand that you have to step out of yourself. And when you're doing that, you're dropping your own self-awareness and self-identity to look at [other points of view]. And then you're able to feel like, 'Wow, okay, I see where I've been doing this and this. Now it makes some more sense to me.' . . . [That process] changed my attitude toward my behavior, and it made me more receptive to listening."

Whack-A-Mole

Despite the positive testimony of soldiers like Pennington, Sharon Sloane acknowledges that there is no silver bullet for combating the wear and tear on the armed forces after more than a decade of war. A game may provide a temporary solution to a problem, but soon enough a new, and perhaps more severe, version of that problem will crop up elsewhere. This is the military's own continuous version of *Whack-A-Mole*—new issues keep springing up and extending themselves faster than the military can (or will) deal with them.

Playing a game like *The War Inside* obviously doesn't cure anyone; nor is it intended to. It is intended, like many of WILL's games, as a bridge *toward* treatment, or toward a change in personal behavior or treatment. The idea that games can play a role in convincing someone to seek treatment, or to recognize his or her own behaviors as unhealthy, is slowly gaining traction throughout the military health-care community at large.

The military is also turning to games for treatment of mental health issues, however. As we are about to see, one of the most promising therapeutic tools for treating veterans' PTSD comes, perhaps surprisingly, in the form of a modified first-person shooter game. But its purpose could not be further from the commercial entertainment of its blockbuster brethren.

CHAPTER 7

The Aftermath: Medical Virtual Reality and the Treatment of Trauma

On September 11, 2001, Jerry Della Salla was a thirty-one-year-old struggling actor who had spent the past decade working in independent films, offering private acting lessons, and living what he calls the "romantic New York artist life." But on the morning of September 12, devastated by the attacks on the city where he'd lived since his freshman year at NYU, he put all that aside. Seized by a desire to become a fireman, he began traveling to firehouses all over Manhattan to find out if anyone would take him. The officials he met all told him the same thing: he was too old. For pension-related reasons, the city wouldn't let anyone older than thirty take the test to join the department.

Della Salla soon decided that the military represented his best option for serving his country. After repeated discussions with a recruiter in Harlem, he signed his papers, entered basic training, and was attached to the army's 306th Military Police Battalion out of Uniondale, New Jersey. There he spent the next two years awaiting orders to deploy. In April 2004, as news broke of the prisoner abuse scandal at Abu Ghraib in Iraq, his unit finally received notice: they would be heading to the military base where the prison was located to relieve the scandal-plagued units and help "restore the integrity" of the U.S. Army in Iraq.

After spending the winter months at Fort Dix — Della Salla and his unit couldn't understand how training in frigid weather in the woods of New Jersey would prepare them for desert warfare in Iraq — the troops of the 306th MP arrived at their destination, the now notorious Forward Operating Base Abu Ghraib, twenty miles west of downtown Baghdad. The soldiers realized immediately that they had walked into what Della Salla calls "a nightmare." "We had entered a hornet's nest of media hype, military shame, and a new motivation for the insurgency," he says. In the aftermath of the revelations of abuse, the base had become a focus of the Iraqi resistance, subjected to almost daily attacks. The pace was relentless. "Everything and everyone outside the wire was trying to come in and take us out," Della Salla says.

The focus of the 306th's mission was "detention operations." Della Salla and his fellow soldiers were in charge of Camp Redemption, a large internment center filled with tents that held twenty-five prisoners each. For fourteen hours a day, the soldiers acted, Della Salla says, like "den mothers," dispensing meals and providing health care. Each soldier generally ran two tents, where he or she interacted with the tent chiefs (primarily imams, or religious leaders), who dispensed the soldier's orders to their tent mates. As a result of the prisoner abuse scandal, the detainees' rights were listed on cards that the soldiers carried at all times. Still, riots inside the tents constantly occurred. For Della Salla, the stress of his work as a "zone dog," combined with the continual insurgent attacks on the base, was nearly unbearable.

Whenever the prison itself became too crowded, Della Salla's unit received orders to "decongest" it, which involved transferring prisoners to other locations. These so-called ConAir operations included running high-risk convoys to the nearest airstrip. Despite the inherent danger, Della Salla found himself wishing daily for one of these missions, because they allowed him to escape the pressure-cooker atmosphere of the base. Anything was better than managing the tents.

One night Della Salla had just finished dinner when he struck up a conversation outside the soldiers' living area with an NCO named Caruso. As they talked, Caruso's face abruptly turned red. Della Salla had his back to the base's walls; it took him a moment to grasp what Caruso was looking at. As he did, Caruso screamed "Incoming!" and grabbed Della Salla and threw him into the living area. At that moment — just after 7 p.m. Baghdad time — a barrage of mortar and rocket fire came streaming over the base's walls. So began the worst three hours of Della Salla's deployment.

Now known as the Battle of Abu Ghraib, the night of April 2, 2005, represented an enormous, highly coordinated insurgent assault on American forces at the base. Small arms, grenades, and even two vehicle-borne IEDs were part of the attack. At the same time, the airstrips in Camp Victory and Fallujah were paralyzed by enemy shelling. A full hour passed before Della Salla's unit and the other battalions at Abu Ghraib received close air support. The insurgents had mined the roads to the west and east, effectively blocking any attempts to send additional ground troops and supplies to support the soldiers on the base. Supplies dipped so low that U.S. forces received orders to fix bayonets for probable hand-to-hand combat.

As the hours ticked by, Della Salla's unit struggled to maintain control over the thousands of prisoners in Camp Redemption. Della Salla also ran medical supplies through an outside quad, where he could hear the enemy rounds getting louder and louder as the insurgents, adjusting their fire, moved ever closer to the base. (The increasing volume of the rounds is a memory that haunts him still.) At one point Della Salla was moving through the quad when a rocket-propelled grenade landed about one hundred yards away, just on the other side of the Jersey barrier, the huge cement slab that marked the outside wall. Though the barrier saved Della Salla's life, the shock waves of the RPG sent him sprawling to the ground. He felt as if a linebacker had tackled him. Still, his adrenaline was running so high that hours passed before he realized he'd been injured by the blast. He was one of dozens of American troops wounded in the battle.

After three hours of heavy fighting, the insurgents finally ended their assault. But any sense of security Della Salla had felt on the base was gone for good.

When he returned to the States after his tour in Iraq, Della Salla moved to the 78th Training Support Division, the unit he had trained under when he was with the MPs. At home he started fighting with his family and friends, and he felt riddled with anxiety, though it took him some time to grasp this, because he had spent the past twelve months in such a state of high tension.

Whenever Della Salla had to report to work with his new unit, he would get in his car and drive from New York City to Secaucus, New Jersey. As he traveled along the highways and beneath the overpasses, his heart would begin to race, he would grow short of breath, and he would find himself gripped by the overpowering fear that he was trapped. Eventually he would become so lightheaded that he'd have to pull to the side of the road until the worst of the symptoms had passed. These panic attacks confused him. He knew he was driving in a car, not in a fortified Humvee, and that he was in New Jersey, not in some convoy on the jam-packed roads outside Baghdad. But this awareness made little difference. Particular triggers — a stretch of highway, an overpass — would set him reeling every time.

Several months after his return, Della Salla began seeing a therapist named Michael Kramer at the VA hospital in Manhattan. At first he and Kramer followed the traditional therapy model: they sat and talked. With Kramer's guidance, Della Salla described his upbringing, his relationships with his family and friends, his experiences at Abu Ghraib, his difficulty in adjusting to life back in the States, and the memories of Iraq that continued to haunt him. He felt the sessions were useful; he had a number of issues that Kramer helped him to express and analyze. At the same time, he continued to feel overwhelming fear, anxiety, and helplessness. He felt powerless to control what he and Kramer had identified as post-traumatic stress disorder.

In 2007, about a year after their first meeting, Kramer asked Della

Salla if he would like to try a new program called Virtual Iraq as part of his treatment. The program combined a virtual-reality/video game component with traditional exposure therapy. Though he trusted Kramer, Della Salla was initially reluctant. "This can't work," he told Kramer. "It's a video game. If I want to use anything virtual, I might as well go to the actual videos I shot in Iraq. That's *real* shit."

Still, Della Salla's curiosity eventually got the better of him, and he agreed to become the first person from the Manhattan VA hospital to use Virtual Iraq. (Kramer, for his part, had been battling the VA bureaucracy to get them to fund the program.) In so doing, Della Salla became one of thirty-five initial active-duty soldiers and veterans to experience what may be the most promising new method for treating psychological trauma.

Virtual-Reality Exposure Therapy

Now operating at dozens of sites across the country, Virtual Iraq and its more recent counterpart, Virtual Afghanistan, are the most widely used virtual-reality exposure therapy (VRET) treatment programs in America. The treatment is a variation on traditional exposure therapy, which itself derives from classic conditioning, in the manner of Pavlov and his dogs. The idea is that by reenacting a traumatic experience or confronting an irrational fear under controlled conditions and then by gradually increasing the intensity of the experience in the context of a safe, therapeutic environment, a patient will become habituated to that experience or fear. The trauma will not disappear, but it will become manageable.

Exposure therapy has a long track record of success in helping people deal with phobias, but it can be prohibitively expensive or even dangerous to conduct on-site. The standard practice, then, is imaginal exposure therapy, in which a therapist repeatedly guides the patient through an imaginary reconstruction of the feared experience. However, this technique demands that the traumatized person vividly recall terrifying experiences, something that PTSD sufferers, given the nature of their condition, are often unable or unwilling to do.

Virtual-reality exposure therapy, made possible by recent technical advances in computing speed, graphics rendering, artificial intelligence, and tracking and interface technology, is a potential solution to those problems. Wearing a head-mounted display (a helmet with goggles and earphones), patients are placed in immersive interactive environments designed to represent their traumatic memories, ensuring that they can confront their experiences without having to conjure up the memories themselves. Since its origins in the early 1990s, virtual-reality exposure therapy has proven remarkably effective at treating anxiety disorders; initial studies indicate a cure rate of between 70 and 90 percent. The Department of Defense has been the largest funder of research, with Virtual Iraq/Afghanistan almost solely responsible for bringing this form of therapy to the attention of the larger psychiatric community. If we ask what effects the military's use of video games will have on society at large, one of the areas of greatest influence will be in the field of mental health. In this sense, the most important video game–related legacy of these wars may have nothing to do with preparing for war at all but be concerned with treating war's aftermath.

With Virtual Iraq, patients are immersed in a variety of settings and scenarios taken from the military-funded commercial video game *Full Spectrum Warrior.* It includes a twenty-four-block city scenario, which features both crowded and desolate streets, a marketplace, empty lots, checkpoints, vehicles (parked and moving), mosques, and a variety of buildings. Users can walk the streets alone or accompanied by computer-animated soldiers; they can enter buildings or climb onto rooftops. It also includes a desert road scenario in which users ride in a Humvee past other vehicles, checkpoints, debris and wreckage, and buildings in various states of disrepair. Users can sit in various parts of the cab or in the exposed turret, as gunners do. (Virtual Afghanistan has its own set of backgrounds, buildings, and other relevant details.) When I tried Virtual Iraq, I was struck by how intense the experience was, though not because of the graphics per se. Yes, the program looked like a video game, but the head-mounted display and other supporting elements combined to make it an unnervingly vivid environment.

A typical course of treatment with Virtual Iraq/Afghanistan consists of ten fifty-minute sessions. The first session includes an intake interview and an overview of the program; the second session includes education about exposure therapy in general and the Subjective Units of Distress Scale (SUDS), which the patient uses to communicate his current level of discomfort to the clinician. The patient then engages in imaginal exposure therapy. During the third session, the patient experiences the virtual world without recounting his traumatic experiences. In the following six sessions, the patient re-creates his traumatic experience while in the virtual world, narrating it with increasing detail and intensity.

Using a so-called Wizard of Oz control pad, clinicians can select the setting for the virtual experience and customize the atmospheric conditions, time of day (night vision is available), ambient sound (such as traffic, wind, or a call to prayer), and even scents (burning rubber, garbage, body odor, cooking spices, gunpowder). They can also introduce IEDs, car bombs, and gunfire. The goal is to re-create the patient's original traumatic experience and then gradually ratchet up the intensity. The scenario doesn't have to match the original experience; it simply has to include similar stressors, such as an exploding IED or a crowd of hostile strangers. Patients control their virtual behavior by manipulating a simulated M4 gun. The weapon cannot be fired; it is intended solely as a mood-setting device.

Throughout a given session, clinicians are in constant contact with the patient, and they can view what the patient is experiencing on their control screens. On the basis of the patient's self-reported SUDS score and the physiological data available (heart rate, galvanic skin response, respiration), clinicians can determine how fast or slow to go. They also ask questions, prompt feedback, and offer support during stressful periods. Each individual session, and the course of treatment as a whole, unfolds at the patient's pace.

Origins of the Game

Virtual Iraq/Afghanistan is the brainchild of Dr. Albert "Skip" Rizzo, a clinical psychologist and associate director for medical virtual reality at the army-funded, USC-affiliated Institute for Creative Technologies in Los Angeles (the original game institute founded by Mike Zyda). Along with his Virtual Iraq/Afghanistan colleagues Barbara Rothbaum, a professor of psychiatry at Emory University, and JoAnn DiFede, director of the Program for Anxiety and Traumatic Stress Studies at Weill Cornell Medical College, Rizzo is one of a small but increasingly influential group of researchers working in the discipline of virtual-reality exposure therapy.

Rizzo presents a striking contrast to many of his military and civilian counterparts. A long-haired, motorcycle-riding, foulmouthed, friendly bear of a man, he prefers leather jackets, jeans, and Harley-Davidson T-shirts to his male colleagues' wardrobes of button-down shirts and khaki pants. He is a passionate rugby player, and his nose, as *The New Yorker*'s Sue Halpern once wrote, "looks like it has met a boot or two." That face is frequently wreathed in cigarette smoke from one of his many daily Marlboros, a habit he picked up, ironically enough, when he ran an antismoking clinic early in his career. Yet beneath his rough surface and laid-back attitude lies a compassionate, driven man who has dedicated his career thus far (he is in his late fifties) to treating traumatic brain injuries and placing new technologies at the service of behavioral health care.

The origins of Virtual Iraq/Afghanistan can be traced to the early 1990s, when Rizzo worked as a cognitive-rehabilitation therapist at a hospital in Costa Mesa, California. Because of his focus on traumatic brain injuries, most of his patients were young men, the population most associated with high-risk behaviors, whether driving drunk or being members of a gang. Patients would come in four times a week for four-hour sessions; during breaks, they would wander outside and rest on the lawn. Rizzo noticed that one of his patients, Tim, a man of about twenty, would spend his breaks sitting under a tree, intently playing with a handheld device that Rizzo had never seen before. One day he asked Tim what he was doing. "And he showed me this Game Boy—it's a new thing, Game Boy," Rizzo says, "and he's playing *Tetris.* And I watched him play, and he was like a friggin' *Tetris* warlord. Here's a kid that was very difficult to motivate for more than ten or fifteen minutes on any one particular cognitive retraining exercise, but here he was, focused and working and getting better at [the game]." Soon it seemed to Rizzo as if all his young male patients were playing on Game Boys on their breaks.

The experience was an eye-opener. Rizzo had been using Apple II-E software for cognitive training, but the software was primitive and not particularly engaging for patients. He felt that computer-based approaches offered a great deal of potential for his work, but it wasn't until he saw those approaches in a gaming context that he realized how that potential might be fulfilled.

Not long afterward, Rizzo got a Nintendo NES loaded with *SimCity.* He became fascinated by the game, and he realized that *SimCity* highlighted everything clinicians refer to as "executive functioning," which refers to the integration of all one's cognitive functions for goal-directed behavior. Aside from learning the interface, players were required to develop and implement a strategy, monitor their performance, and revise and update that strategy—"all the kinds of things the brain typically does in complex everyday situations," Rizzo says. He took *SimCity* in to his patients, and they loved it; they spent hours focused on their game play.

While Rizzo was becoming interested in gaming, virtual reality was seeping into the public consciousness. He began to consider returning to academia, where he could chart a path for harnessing these new technologies in the service of clinical practice. He was not by nature a technological person, but he saw how virtual reality and gaming could be an ideal match for exposure therapy. "That was a no-brainer right out of the gate," he says. In 1995 he took a postdoctoral position at USC's Alzheimer Disease Research Center, where he drew on his years of clinical experience with Alzheimer's patients and patients with traumatic brain injuries. Still, he admits, "the real mission was to make friends with the folks who were in computer science and get access to a foothold."

Over the next several years, Rizzo poured himself into designing virtual-reality systems that could be used in clinical settings. After his postdoctoral work, he took a position at USC's Integrated Media Systems Center, where he and a programmer began running studies involving virtual reality.

Rizzo was still working at the center in March 2003, when the United States invaded Iraq. "I'm watching all this stuff, 'mission accomplished,' all this horseshit," he says, "and I'm thinking, *You know, they're talking about nation building — this isn't going to be a cakewalk.*" He realized that the war would probably produce a generation of traumatized veterans. To avoid another post-Vietnam situation, he believed the health-care profession needed to work quickly on finding effective and accessible treatments for PTSD.

While preparing for a talk on the subject, Rizzo visited the website for the Institute of Creative Technologies to see if he could find any relevant material. There he found a clip of the institute's newly developed video game *Full Spectrum Warrior.** "When I saw [the game]," Rizzo says, "it looked *just* like Iraq. In my mind, anyway — I'd never been to Iraq. But it had that Middle Eastern look to it. And I thought,

---

* The game remains the best-known — and most controversial — project to emerge from the ICT. A big success commercially, *Full Spectrum Warrior* ended up being an embarrassment for both the army and the ICT, as it proved utterly ineffective as a training tool. In essence, the army paid millions of dollars to produce a game that it couldn't use and that it wasn't allowed to profit from. Meanwhile, the game's commercial partner, THQ — which hadn't spent its own money — raked in over $50 million in sales.

Why can't we just take this content and modify it and make it into a therapy tool?"

Rizzo contacted Jarrell Pair, an ICT researcher who in 1997 had programmed Virtual Vietnam, the first virtual-reality application for PTSD. Developed by researchers in Atlanta, Virtual Vietnam immersed users in one of two settings, a jungle clearing or a Huey helicopter, with a limited number of customizable details. Though the graphics were crude, the program seemed effective. A case study of a fifty-year-old veteran with treatment-resistant PTSD yielded promising results, as did a subsequent small-sample controlled study. But the results were never followed up, and the project soon came to an end.

Rizzo and Pair began working on a prototype of Virtual Iraq, which they finished in early 2004 and which used graphics from *Full Spectrum Warrior* to depict an Iraqi market street. When Rizzo applied for money to develop the project further, however, he was turned down. He continued to knock on doors until July, when the *New England Journal of Medicine* published a paper by epidemiological researcher Charles Hoge and his colleagues at Walter Reed Military Medical Center. The article outlined for the first time the high incidence of PTSD among soldiers in Iraq and Afghanistan. Hoge's research sounded the alarm for the military and the public alike.

A few weeks afterward, Rizzo received a phone call from Russell Shilling of the Office of Naval Research. Shilling was a former member of Mike Zyda's MOVES team at the Naval Postgraduate School; he had been both sound designer and principal investigator on *America's Army*. Given his longstanding interest in advanced technologies, he needed no prodding from Rizzo to grasp Virtual Iraq's potential. Though the Office of Naval Research did not traditionally fund clinical research, Shilling managed to get his hands on some money, and in March 2005 he delivered the funding Rizzo needed to continue his work.

Rizzo began conducting feedback sessions, in Iraq and at the Naval Medical Center in San Diego, with psychologists, military personnel, and veterans. Hundreds of soldiers and veterans also provided feedback on the USC campus. Following these sessions came a clinical trial featuring active-duty soldiers. At the end of the trial, 45 percent of subjects no longer had PTSD, while an additional 17 percent showed improvement in their symptoms — a better-than-average rate in treating post-traumatic stress disorder.

Rizzo envisioned Virtual Iraq as benefiting clinicians as well as patients. "To be a good exposure therapist, you've got to have some kind of imaginative skill," he says. "You have to help guide and be comfortable with some of the hard things that people recall from the imaginal narratives. With virtual reality, you've got a very structured tool for pacing the exposure. So if you're not such a great imaginative exposure therapist, you can become a very good one using this tool."

The third benefit of the program, Rizzo felt, was that it could make mental health treatment less stigmatizing for soldiers. "That's always been one of the pitches. [Maybe they'll] say, 'Oh, it's kind of like a video game; I'll give it a try.' It's still hard therapy, but it's in that context. Soldiers that have gone through the treatment all say the same thing: 'I'm glad they put it in this format. It's better than having a doctor try to pull it out of you.'" A 2008 survey of more than three hundred active-duty service members showed that one in five people who categorically refused to try traditional therapy would be willing to try Virtual Iraq.

The Monster in a Box

When Jerry Della Salla began using Virtual Iraq at the Manhattan VA hospital, he and his therapist, Michael Kramer, were embarking on a relatively uncharted course. Still, the shape of their sessions matched what has become the program's standard method for use. Based on his previous discussions with Della Salla, Kramer ordered up a convoy operation as the basic scenario for Della Salla to work through. Because the treatment involves gradually raising the patient's stress level, over the course of several sessions Kramer added details that increased both the realism and the intensity of Della Salla's experience. He added explosions and gridlocked roadways and bloody soldiers and civilians. Using the program's scent machine, he mixed the smell of diesel from a Humvee with the odor that emanates from a fired weapon. He added details from the Battle of Abu Ghraib, particularly the deafening whistle of incoming mortars and RPGs, two of the sounds that continued to haunt Della Salla. As these details were added, Della Salla narrated his experiences over and over again, until the combination of sights, smells, sounds, and his own memories made him want to tear off his head-mounted display and run from the room.

When Della Salla reached these points of ultimate stress — usually about thirty minutes into a session — Kramer would feed him strategies for working through his ragged emotions. He would then coax Della Salla through yet another convoy scenario, urging him not to stop midway. (Because the patient controls the joystick, he controls the speed and length of a scenario.) According to Della Salla, "As long as you're communicating in the process and allowing [the therapist] to bring you down from any sudden moments of panic, that allows you to see what's going on with your body, with yourself, and then to have the strength and the ability to just slow it down."

As the weeks went by, Della Salla found himself gaining confidence in his ability to make it to the end of a scenario. The difference was subtle at first, but over time he began to feel less as if his emotions were controlling him and more as if he could control his emotions.

This shift points to a key issue with PTSD, as well as with its treatment. Patients like Della Salla often feel at the mercy of their own bodies and minds, and the fear generated by this inability to regulate their responses can sometimes be as crippling as their traumatic memories. For Della Salla, learning that he was no longer helpless in the face of his disorder started him on the path to recovery. Recovery is different from a cure, of course. The traumatic memories didn't go away; they simply lost some of their power over him. "PTSD is this monster in a box," Della Salla says. "And you have to respect it enough to know how to control it, because at some point it has control over you. And it's not your fault. Once you make the decision to accept that, you can maybe, hopefully, learn how to put it in its place."

Della Salla says that his discussions with Kramer over the first year of treatment prepared him for what Virtual Iraq offered: a more intense, immersive level of therapy in which he could more easily make progress. The program enabled him to work on issues that he would not have confronted otherwise. Without it, Della Salla says, he wouldn't be where he is now. "What I like to believe is that you do get stronger. You get more confident. That's probably the best word. But you shouldn't let your confidence be the thing that puts your guard down."

"I would be the first person to say that technology doesn't fix anybody," Skip Rizzo admits. "Technology's just a cold tool that can help the therapists do their job better. And that's what I think Virtual Iraq/Afghanistan is. And I think that's what the power of simulation technology is. The human brain has this great propensity to suspend disbelief when we're watching a TV show, or a movie or a play or whatever. Well, simulation technology can do it in a more comprehensive and relevant way, and in a very controlled fashion." When PTSD sufferers begin to gain control over their memories in therapy, they embark on a process that psychologists refer to as "extinction." "They're reliving the memories," Rizzo says, "but there's no real bad thing objectively happening. Yes, there are bad memories, but memories can't objectively hurt you in your current life. They can haunt you a little bit, but that's where the emotional processing, by telling the story, occurs."

According to the *Journal of CyberTherapy and Rehabilitation,* more than fifteen studies entailing diverse populations have shown that virtual-reality exposure therapy enhances traditional cognitive behavioral treatment regimens for PTSD. The journal reports that most studies reveal a treatment success rate of 66 to 90 percent. An issue of *Studies in Health Technology and Informatics,* meanwhile, summarized case studies from a navy-funded project comparing the effects of VRET with the effects of traditional treatment on active-duty navy corpsmen, Seabees, and navy and Marine Corps support personnel.

The results showed that VRET led to measurable reductions in reported symptoms of depression, anxiety, and PTSD.

Most recently, *Military Medicine* reported on a treatment project aimed at developing and testing a method for applying VRET to active-duty service members diagnosed with PTSD. Forty-two service members were enrolled; twenty of them completed treatment. Of those twenty, 75 percent experienced at least a 50 percent reduction in PTSD symptoms and no longer met accepted medical criteria for PTSD after treatment. On average, PTSD scores decreased by 50.4 percent, depression scores by 46.6 percent, and anxiety scores by 36 percent. Analyses showed that statistically significant improvements in PTSD, depression, and anxiety occurred over the course of treatment and were maintained at follow-up.

Crucial to the growth of VRET, of course, is the willingness of clinicians to use it. A recent issue of *Psychiatric Services* presented the results of a study gauging Veterans Health Administration mental health clinicians' perceptions of virtual reality as an assessment tool or a component of exposure therapy. Although the study demonstrated that the use of virtual reality as a therapy was feasible and acceptable to clinicians, it also showed that successful implementation of the technology as an assessment and treatment tool will depend on consideration of the helps and hindrances, bureaucratic and otherwise, in each given setting.

Given the positive track record of the initial studies on VRET and PTSD, the military would do well to remove any barriers to continued implementation and expansion of the therapy. For soldiers like Della Salla, VRET may be key to managing recovery. Marine Master Sergeant Robert Butler, who spent a year in Iraq and returned home with the classic symptoms of PTSD, acknowledges that VRET "is tough. It's tough. Because you've spent so much time trying to avoid thinking about your deployment, and they're dredging up these memories that you tried to avoid at all costs. It's difficult." Still, Sergeant Butler says, "I think it was a great idea for them to put treatment in that [virtual-reality] format. Better probably than just sitting there and having some doctor try to pull the events out of you. You're right there. Boom. Smack. Face-to-face with your worst demons.

"I mean, am I a hundred percent better? No, I wouldn't say I'm a hundred percent better. But I do have my life back. I'm able to do a lot of the things that I did before . . . I'm not running around angry all the time. This treatment was . . . it saved my life, probably. It saved my marriage, for sure. So if you asked me if it works, I would say, 'Yeah, it works.'"

Wars of Innovation: Medical Virtual Reality in the Twenty-First Century

"War sucks . . . but it does drive innovation." So reads the PowerPoint slide with which Rizzo often begins his public presentations. As he points out, these innovations have had a major impact on civilian health care and mental health rehabilitation. Indeed, as much as I have emphasized the military's influence on education, it has had an equal or greater impact on American health care, including the development and growth of the field of psychology. "It's a sad commentary on humanity, but it's a reality," Rizzo says. "The military is out front."

When Rizzo received his initial funding for Virtual Iraq from the Office of Naval Research, he realized that the project had the potential to branch out into areas of health care beyond the treatment of PTSD. He approached Randall Hill, the Institute for Creative Technologies' executive director. "Virtual Iraq could be a good start-off," he told Hill. "I think I'll be able to grow a research program, because the military's going to want more than just PTSD treatment." Hill agreed, and Rizzo officially joined the institute on October 1, 2004. In the intervening years, true to his word, he has developed not only Virtual Iraq/Afghanistan but a series of other projects that use video game technologies in the service of veterans' health care.

Today Rizzo oversees four affiliated labs, known collectively as the ICT MedVR Lab, all of which grew out of Virtual Iraq. He conceptualized and orchestrated these labs, then brought other people in to run them. (As a manager, Rizzo now spends most of his time writing grants and papers and networking.) The Virtual-Reality Psychology Lab focuses on PTSD, stress resilience, and pain distraction work; the Motor Rehabilitation Lab addresses physical rehabilitation; the NeuroSim Lab focuses on neuropsychology; and the Virtual Patient Simulation Lab leverages the ICT's decade of work in developing virtual humans, which Rizzo's team translates into clinical tools.

The most developed of these clinical tools is a Web-based application called *SimCoach,* in which virtual humans act as advisers and sounding boards for members of the military community who may be suffering from such issues as depression, stress, substance abuse, suicidal ideation, brain injuries, and relationship difficulties. It is a pre-intake tool, as opposed to a form of online therapy. One of the primary goals of *SimCoach,* which cost $10 million to develop, is to break down the barriers to care that have traditionally dogged the military — most prominently, the belief among soldiers that they will be judged negatively if they admit to needing help. A recent Mental Health Advisory Team study of soldiers in Afghanistan found that over 50 percent believed that they would be seen as weak if they sought behavioral health care, while 34 percent believed that seeking help would harm their careers. Because users interact with *SimCoach* anonymously, Rizzo and his team are hoping that soldiers will be inclined to use it. Though it features the same basic content as WebMD, the use of virtual characters to relay that content adds a social dynamic that is intended to engage patients and keep them involved as they gather information.

There are other barriers to health care for military personnel, including accessibility and availability — whether there are enough health-care providers in a given location and whether that location can be easily reached by service members in need of help. The American Psychological Association's Presidential Task Force on Military

Deployment Services recently declared itself unable "to find any evidence of a well-coordinated or well-disseminated approach to providing behavioral health care to service members and their families." Because *SimCoach* is a Web-based application, it is available anywhere at any time, which will address at least some issues of accessibility and availability.

One bright October morning, I sat in a sunlit conference room at the Institute for Creative Technologies as Rizzo presented *SimCoach* to a group of faculty members and doctors from USC's Keck School of Medicine and the Los Angeles County Hospital. The meeting was part of a push by Rizzo and his team to promote and expand *SimCoach* and its related applications — by having them taken up first by the broader USC community and then by the nationwide medical community. Rizzo had even dressed up for the occasion: he was wearing black jeans and a black button-down shirt and had pulled his hair back in a ponytail.

As the presentation began, we were introduced to Bill Ford, one of *SimCoach*'s virtual characters. Bill is a gray-haired Vietnam veteran, a white man with a southern accent who talks in a cloyingly folksy manner. Wearing a long-sleeved gray shirt and blue jeans, Bill sits at a wooden table on the back porch of a house, fields and trees spreading into the distance. (*SimCoach* looks like a video game.) With a cup of coffee near at hand, Bill starts things off by saying, "The suits want me to explain to you that I'm not real." Once this detail is out of the way, Bill talks about a soldier friend of his, Jared, who has recently returned from Afghanistan and Iraq and who is having a number of problems now that he's back home. As Bill discusses Jared's various issues — he's depressed, he's fighting with his family, he's having flashbacks — questions appear at the bottom of the screen. "Are you experiencing similar issues?" the first question asks. (Soldiers using *SimCoach* answer this question and similar ones, all of which are designed to provide an outline of what they might be struggling with.) Eventually Bill stops talking about Jared. "All right," he says to the user, "from what you're telling me, it looks like you're

having flashbacks
upset at memories
avoiding things that might trigger memories
Is that right?"

Depending on how the user responds, Bill suggests various resources that might be helpful: articles, websites, video testimonials, support groups, lists of local providers. "Here's a link to people in your area," he might say, and follow that up with "Do you have any idea of what to look for in a counselor?" If the user responds in the negative, then Bill provides additional guidance.

Though Bill Ford is somewhat cheesy, he is also undeniably helpful. The question for Rizzo and his team is whether soldiers will suspend their disbelief when they interact with Bill and the other *SimCoach* characters. (Rizzo is currently running a study to determine the answer to this question.) Previous studies have shown, surprisingly, that graphics are not the most important element when it comes to virtual humans. Instead, character movements (hand gestures, head nods, and so on) are the key to establishing rapport. Studies have also shown that people respond to virtual humans in the same way that they respond to real humans, simply because we don't know any other way to react.

Rizzo's team has also developed virtual humans that will be used to train therapists and social workers in dealing with military populations. USC's School of Social Work features a pioneering concentration in military social work, intended to prepare students for the special needs of the nation's military personnel. At the same meeting at which Rizzo presented *SimCoach,* Patrick Kenny, the director of the ICT's Virtual Patient Simulation Lab, showed off his lab's latest creations, which the School of Social Work will soon employ. Kenny began by describing the limitations of using live actors, who have traditionally been used to train budding social workers. To begin with, he said, live actors require a great deal of training before they can inhabit their roles in a believable fashion. The high rate of turnover among live actors is also a negative, as is their lack of availability on a 24/7 basis. Child actors in particular, Kenny said, can be quite difficult to find.

By way of contrast, Kenny is developing a series of standardized virtual patients who don't suffer from any of these real-world faults. He shows off Alamar Castilla, the lab's most developed virtual patient. Castilla, the story goes, has just returned from his third deployment overseas and is suffering from PTSD. Fidgety, angry, and clearly uncomfortable in a therapeutic setting, Castilla requires a subtle approach on the part of a mental-health-care worker. Kenny shows us a worst-case scenario — the student clinician asks a number of inappropriate questions, including, most glaringly, "So, did you ever kill anyone?" "Fuck this, I don't need this shit," Castilla responds before storming off in a huff. The simulation is hardly perfect: Castilla's mouth movements don't match his words, and his voice is disconcertingly robotic. Because voice recognition software is less than ideal, Castilla and the other virtual humans sometimes offer incorrect responses. At this point, too, the questions that Kenny has designed are exclusively closed-ended, meaning that they are directed to very specific responses; open-ended questions are much more difficult to program. These are all issues that Kenny and his team are still working on.

Later in the day, the other attendees and I were introduced to the *Immersive Naval Officer Training System,* another video-game-looking program that is built around virtual humans. The program is designed for young officers just out of the Naval Academy who must counsel older enlisted personnel about sensitive personal issues as part of their jobs. What happens, for example, when a baby-faced twenty-two-year-old officer has to give advice to a sailor in his thirties who is worried that his wife is about to leave him? We are shown a virtual scenario in which Gunner Cabrillo has just shoved Gunner Thomas in the workplace and has now been summoned to his commanding officer to explain himself. (The user of the program plays the officer.) Cabrillo apologizes for his actions, but he says that Thomas was "talking shit" about his wife by claiming that she was cheating on him. The user must school Cabrillo on how to deal with a situation like this: he should report it to his commanding officer immediately instead of getting physical. This leads to another scenario in which Cabrillo finds out that Gunner Thomas was correct: his wife is in fact cheating on him. The issues get more complex as the program continues.

Following a long day of demonstrations, Rizzo gathered the USC faculty members and doctors in a circle to discuss what they had seen. He was joined by former army Major Thomas "Brett" Talbot, chief scientist of the Armed Forces Simulation Institute for Medicine and the ICT's in-house medical doctor. Talbot had led the demonstrations that day, and he and Rizzo were anxious to know whether their audience would use programs such as *SimCoach* in their own medical facilities. For example, Talbot said, he and Rizzo were working on a proposal that would make their virtual standardized patients available for free to medical educators. Would the attendees be interested in that? Yes, came the response. Talbot explained that he wanted to develop a critical mass of standardized patients, and that required the kind of buy-in he and Rizzo were hoping to get from those in the room. "What the ICT wants from you," he explained, "is the opportunity to connect. Maybe you could apply for your own grants to utilize our technology. Maybe you could share with us your expertise on standardized patients and give us feedback on future applications. Maybe you could connect us with the medical educators you're friendly with around the country."

Talbot clearly felt that he wasn't receiving the kind of energetic response he had hoped for. "We are on the precipice of some major awesome thing with this technology," he exhorted the room. "We want you to join us!"

Rizzo stepped in to continue pushing the argument. "We're building on twelve years of Department of Defense funding," he said. "Now we're at a point where dual-use opportunities are at a premium. You can use the military-funded data we've collected for your own grant applications." He encouraged the people in the room to be leaders in what the ICT wanted to accomplish.

If military-funded innovations in interactive digital training and education spread into the civilian realm, this is how the process will occur. There is no sudden explosive growth here, no concerted Pentagon effort to exert its influence in the civilian arena. There is instead personal contact, a growing network of connections, and the relentless efforts of people such as Rizzo and Talbot to make their work matter on the largest possible scale.

Comprehensive Soldier Fitness

Recently the army unveiled a new program called Comprehensive Soldier Fitness (CSF), described as "a structured, long-term assessment and development program to build the resilience and enhance the performance of every Soldier, Family member and [Department of the Army] civilian." Coming after more than a decade of war in Afghanistan and Iraq, CSF is aimed at developing "balanced, healthy, self-confident Soldiers, families and Army civilians whose resilience enables them to thrive in an era of high operational tempo and persistent conflict." The idea is that in a time of permanent warfare, the army community needs a top-down initiative that can be used before deployment to build individuals' abilities to bear the mental and physical strains that inevitably result from battle. By so doing, the army hopes to stem what has been an explosive increase in war-related social and psychological problems: PTSD, broken families, an exhausted and demoralized workforce. Critics say that the CSF program is designed to turn soldiers and their families into automatons who, against human nature, will remain unaffected by the trauma of warfare.

Though he is fiercely liberal, Skip Rizzo is not one of these critics. He wants to use gaming and simulation technology to build resilience-training systems that will be more effective than the traditional death-by-PowerPoint briefings that teach soldiers how to deal with their stress. "We believe we can better prepare people for the emotional challenges they're going to face when they go off to combat by using very structured, guided, story-based simulations," he says. Specifically, he envisions a series of five- to ten-minute-long episodes like those in *Band of Brothers* that would have the benefit of being, like a video game, fully immersive. "You'd be right in the Humvee with a bunch of digital characters, and they'd be talking like in the helicopter ride in the first *Predator* movie, when they're taking off and everyone's telling jokes and shit," he explains. "But you'd build that narrative, that story, so that people would start to bond with those digital characters the way they would bond with characters in a movie. And at the end of each episode, something would go south. It might involve getting attacked. Or having to handle human remains. Seeing a civilian getting killed and not being able to stop it. Seeing one of your comrades get killed or grievously wounded. Accidentally killing somebody. And at that moment, in would walk a virtual mentor to guide you through: 'How are you appraising this? What was your experience before? What's your hypothesis about why this happened or what your role or responsibility was?'"

Rizzo has received seed money from the military to build three of these episodes, each of which will incorporate a wide variety of physiological measures of the user's stress reaction. "I envision this as being a thirty-episode training program that you start after you get out of boot camp, and that you do in your own barracks," he says. "You go through the process and you have to take a test after, and so on and so forth. Point being, here you have a simulation at an entry point. And maybe you can train people how to be better copers by putting them in that context."

You Have to Have a Champion

Early one evening, at a dingy bar on Marina Del Rey's main thoroughfare, I met with Rizzo and Talbot to discuss their work further. As excited as he was by the MedVR Lab's various initiatives, Rizzo was anxious to run additional user studies so that he could gauge how soldiers reacted to them. His research over the past several years had indicated that people respond most to a story — that a compelling narrative is the best way to genuinely engage people and teach them. Virtual Iraq/Afghanistan had received funding for repeated randomized trials and would eventually be implemented at every VA hospital in the country; Rizzo hoped that *SimCoach* and his team's other initiatives would receive a similarly positive response.

"This shit is flyin'," Rizzo said of his work, but where would it end up? Would the Pentagon actually dedicate the resources that were required to provide the military community with effective behavioral health care? In his years of dealing with the military, he had come to realize that "you have to have a champion, someone who wants to make change happen, who wants to make change *matter* for people." In his case, that champion was Russell Shilling, who not only had provided the initial funding for Virtual Iraq but had funded *SimCoach* through his current position at DARPA. With Virtual Iraq, when the army learned that the Office of Naval Research was funding it, it decided not to join in. These kinds of divisions and rivalries between the services, Rizzo said, inevitably end up hurting the soldiers on the ground, not the leaders in their finely appointed offices back home.

Comprehensive Soldier Fitness is supposedly different — it is a top-down army initiative that acknowledges the past decade's wear and tear on the force. When I suggested to Rizzo and Talbot that the program would inoculate military personnel against rational human responses to the pain and suffering of war, Talbot vigorously disagreed. "Think about these young soldiers who enter the military," he said. "They're joining to escape their shitty lives — they have no idea what to expect. Then they deploy, and all of a sudden they're experiencing hell on earth. When they return home, their spouses leave them. They have no coping strategies. Stress resilience isn't designed to make them *robots,* it's designed to make them *human* — to make them willing to engage with their emotions and seek help. How are nineteen-year-olds supposed to deal with seeing their friends killed, with seeing babies blown up? These are normal people dealing with abnormal stress." Rizzo said that critics should be glad that a liberal like him is working with the military and helping to shape what it is doing. He is, he said, keeping watch.

Earlier in the day, I'd been in a meeting with Rizzo and a group of his researchers as they discussed turning *SimCoach* into an assessment tool. They had recently received a $950,000 grant to adapt *SimCoach* for use in an annual assessment program in which all military health-care providers are required to participate. A few days later, the army would be visiting ICT to see how the project was progressing. Rizzo and his team wanted to sell the army on their vision while taking care to manage the army's expectations of what could be accomplished. "This is a grand opportunity because it puts us and *SimCoach* in front of all these military professionals," Rizzo told the group. "It's a key opportunity for dissemination. We need to engage in shameless self-promotion."

Later, as the meeting reached its halfway point, Rizzo was struck by a thought: despite the many stresses of his job, the innumerable daily headaches, was this, for him, the most productive time of his life? He had around him a team of people whom he believed in, and who believed in him and were doing this work not for the money — compared to the corporate world, there wasn't much — but because they believed in it. He found himself caught up in an imagined moment of future nostalgia. He tried to stay in that moment, to let it fully absorb him. Then he pulled himself out of his reverie and plunged back into the matter at hand.

CHAPTER 8

Conclusion: *America's Army* Invades Our Classrooms

IN 2008, THE ARMY'S 3rd Recruiting Brigade of Ohio confronted a dilemma: how could it inject the army's presence into Ohio's public schools when so many educators and parents were opposed to recruiting on school grounds? This is an issue that faces recruiters across the country, but alone among them, the 3rd Recruiting Brigade hit on a novel solution: it would partner with Project Lead the Way, a nonprofit educational content provider, to "promote student interest in the engineering and technical fields" by using *America's Army* in high school classrooms across the state. Visuals and scenarios from the game would be repurposed for modules in a variety of courses. Because Project Lead the Way's curriculum is preapproved in all fifty states, the *America's Army* part of the curriculum would automatically be permitted in schools nationwide.

The gambit worked, and following a successful yearlong pilot backed by the Ohio Department of Education, Project Lead the Way began giving its *America's Army* learning modules free of charge to public schools around the country. Richard Grimsley, Project Lead the

Way's vice president for programs, told me that the game is currently used in all fifty states, in upward of two thousand schools.

The *America's Army* employees in charge of the Project Lead the Way collaboration — Craig Eichelkraut, the project head, and Catherine Summers, its education/training liaison — were quite open in confirming the army's motives. When I met with them at Redstone Arsenal in Huntsville, Alabama, Eichelkraut said that the army's goal was not only to embed a positive image of the military in the public schools but also "to have a way to get into the schools [in the first place]. Because right now, in a lot of schools, if you're a recruiter and you come to the school, they just do not want you even to talk to the students." Bundling the game with Project Lead the Way means, Eichelkraut said, that the army "doesn't have to fight with every single school and every single school board." He also told me why the army chose the course "Principles of Engineering" as its first module: "Principles of Engineering goes into one of the classes that every student has to take, so we felt that this would be the one that would hit the most students."

The army and Project Lead the Way's shared interest is in developing school-age populations with skills in science, technology, engineering, and mathematics. From the army's perspective, these so-called STEM skills are essential to the success of the current and future fighting force; for Project Lead the Way, these are the skills that will keep our country's workforce at the forefront of the global economy. As the two organizations collaborated, they disagreed primarily on the size of the explosions that would be allowed in the game. Equally important, Project Lead the Way "wanted to make sure that the kids were understanding the mathematics behind projectile devices, those kinds of things," Richard Grimsley told me, "as opposed to just blowing stuff up. That was a big challenge for [the army]."

The partnership with Project Lead the Way is just one means by which *America's Army* is making its presence known in our schools. Now that Casey Wardynski is Huntsville's school superintendent, the

America's Army team at Redstone Arsenal is creating applications of the game for the Huntsville curriculum. These applications will be delivered to the classroom via iPads and iPhones; they will also be recycled into the game's public recruiting and private training versions. Wardynski has partnered with the U.S. Army's Cyber Command to restructure the curriculum of Huntsville schools to focus on the skills required to wage and defend against cyberwarfare effectively. Beginning in 2013, the city's middle schools and high schools incorporated a program targeted at developing students into skilled "cyberwarriors or cyberdefenders"; the plan is for these students to enter the army immediately after high school. (The army's Cyber Command is providing the appropriate curriculum, along with soldier-mentors.) When they complete their military service, these Huntsville graduates will return to the city to work for one of the many defense contractors in the area. They will also have the opportunity to enroll in college. Ideally, for Wardynski, this model will be replicated by school districts across the country.

While many educators are opposed to a military presence in their classrooms, Susan Tave Zelman, Ohio's state superintendent of public instruction, is not. "When we were approached by the U.S. Army," Zelman says, "we realized the great opportunity this project represented for engaging students in a learning environment that excites them ... This marks a real shift in the education paradigm to utilizing a technology platform that students are familiar with and enjoy!" Equally keen on the military–public schools alliance are Brenda Welburn, executive director of the National Association of State Boards of Education (NASBE), and the organization's past president, Brad Bryant. Announcing the army-sponsored "Building Strong Futures Together" conference, held at Fort Jackson, South Carolina, Welburn declared that the conference would "stimulate, and sustain, dialogue with one of our nation's largest employers of our public school system." Bryant, in an open letter to NASBE members, enthused that attendees would "have the opportunity to test [the army's] weapons simulation, ... do humvee rollover simulation, and overcome our fear of height [*sic*] with the Tower exercise. Can we say high school hands on learning!"

To school officials who collaborate with the military, the relationship makes perfect sense: the Pentagon has money and our public schools are starved for funds. This logic is driving the increasing corporate presence in education — the brand names that pop up regularly in textbooks, or the exclusive vending deals that schools are signing with Coke and Pepsi. But the military connection takes this process one step further: it delivers cutting-edge learning tools into the classroom, creating curricular areas where the immersion in, say, army-branded virtual worlds may define the learning experience itself.

The army's primary reason for entering our schools is, as I have said, recruiting. But more specifically, as the Project Lead the Way example shows, it is recruiting young people with certain skills. Conservative and liberal commentators share a rare point of agreement in saying that only by vastly increasing and improving STEM proficiency among our schoolchildren will America's future be as glorious as its past. The Obama administration agrees: it is pouring hundreds of millions of dollars into its Educate to Innovate campaign, a pro-STEM initiative that, in the president's words, is dedicated to "reaffirming and strengthening America's role as the world's engine of scientific discovery and technological innovation." And while the conservative Heritage Foundation has sharply criticized the Obama team's efforts, it is not, as might be expected, for the heavy use of federal funds, but because the foundation wants state and local policymakers to have more flexibility in determining how to spend those funds.

The Heritage Foundation's report states explicitly that a STEM-educated workforce is essential to our nation's "global competitiveness" and "national security," while bemoaning the fact that, with regard to STEM, American students "are now outperformed by students from Liechtenstein, Slovenia, Estonia, and Hungary." The foundation approvingly quotes the 1983 study "A Nation at Risk," the National Commission on Education's notably hysterical report on America's fallen status:

> Our nation is at risk. Our once unchallenged preeminence in commerce, industry, science, and technological innovation is being overtaken by competitors throughout the world . . . What was unimaginable a generation ago has begun to occur—others are matching and surpassing our educational attainments. If an unfriendly foreign power had attempted to impose on America the mediocre educational performance that exists today, we might well have viewed it as an act of war.

Claims that education is essential to national security are as common today as they were in the World War II era, when entire school programs were given over to propaganda films and texts that impressed the virtues of patriotism and the bravery of the armed forces on receptive children nationwide. The contemporary hue and cry that American schools have lost their competitive edge to those of China and other countries resembles the panic of the 1980s, when our nation's pundits knitted their collective brows over the ascendance of Japanese industry and innovation. The use of video games gives this old argument a uniquely twenty-first-century twist. The Obama administration, for example, recently established the National STEM Video Game Challenge, which it describes as "a multi-year competition whose goal is to motivate interest in STEM learning among America's youth by tapping into students' natural passion for playing and making video games." In an unprecedented move, the White House also hired a video game czar to craft its national video games policy. Dr. Constance Steinkuehler, a professor of education at the University of Wisconsin–Madison, took on the role of senior policy analyst in the White House's Office of Science and Technology Policy, where she guided the administration's strategy for promoting and harnessing the power of serious games in the fields of education, civic engagement, health, and the environment, among others. Steinkuehler was charged with building new links between government, academia, and the private sector.

The more this occurs, the greater the military's influence on our nation's approach to twenty-first-century learning will be. After all, the military has long been at the forefront of the serious-games movement, as Steinkuehler herself told me years ago, when she sat on my dissertation committee. Not only was *America's Army* the first — and arguably still the best-known — blockbuster serious game, but the military is employing games multilaterally and to a much larger extent than any other entity, as we've seen. Moreover, we are poised to experience an explosive rate of growth for the nonmilitary uses of video games as learning tools. The use of games could dramatically reshape the nature of learning and education in the decades to come. The *New York Times Magazine* reports that a growing number of influential education professionals believe that school should be remodeled to resemble a good video game more closely — meaning that it should emphasize active, immersive, situated learning. In states such as North Carolina, Virginia, Nevada, Arizona, Texas, Colorado, and Massachusetts, video game–based learning has already taken root in selected schools. New York City recently opened Quest2Learn, the nation's first public school to feature a curriculum based entirely on the principles of good game design. (The school's founder and executive director, Katie Salen, is also a video game designer.)

By using video games to teach, the military continues its long tradition of defining new modes of instruction. And yet it is *because* of video games' educational potential that we need to question the ends to which this potential is being directed. Speaking of the army's game efforts, Michael Macedonia has observed that "the big challenge isn't getting the technology right. We're almost there. The challenge is: Do we have the right story? Does it map to reality? Are we teaching the right thing?" Similarly, when it comes to the military influencing our schools, how will we determine what "the right story" is? And how will video games be used to shape this decision?

Like those in the serious-games movement, I believe that video games have a broad range of positive learning applications. So what might be the possible dangers of the scenario I have just outlined? What is troubling about the military's use of video games influencing our schools, or about the military's *actual* games being adapted for our schools? One issue, as Douglas Noble points out, is that historically, the joining together of education and advanced technology has not "been driven . . . by the real and pressing needs of education." Put another way, what the military wants and needs from learning is not necessarily what educators and students want and need from learning, and yet the military has for decades shaped American education to suit its needs. We can see this in the use of computers in schools; as we saw in Chapter 2, military research has been the primary force driving the creation and use of educational technology since World War I. This military imprint has helped to define what Noble calls "the purposes and the 'products' of education" in our high-tech society — a society that, as we have seen, is based on technologies derived in large part from military R&D.

Another concern relates to Noble's contention that individual technologies reflect "the particular goals of those parties or institutions with the resources and the power to determine the shape of the technology." The possible problem with this is evident in Princeton historian Paul Starr's argument that choices related to technological architectures can represent "politics by other means, under the cover of technical necessity." Nor, Starr writes, does the issue stop there. Once a technology has been developed, it continues throughout its lifetime to assume a shape and character broadly reflective of its original context. If, as with *America's Army,* versions of the military's games are used in our schools, the ideological ramifications are more straightforward. As J. C. Herz writes in *Joystick Nation,* "The one thing . . . political simulations all share is the insistence that you, the player, are in control . . .And it's really easy to get that impression, because you're taking such an active role, and because the system works in this pseudomechanical way that seems transparent. But of course, this transparency is sim's first and greatest illusion. Sim is not neutral . . . Every sim has a set of embedded biases and assumptions."

In Herz's estimation, the interaction between player and game amounts to a form of "social contract" in which, at least while the game lasts, the player accepts "the designer's values and assumptions."

This itself is not an issue. The issue is that most players don't realize the nature of this contract, especially if the game is dressed up in bells and whistles and "lavishly produced." This, Herz writes, is "what makes sim so effective at convincing people that certain types of political behavior are appropriate. Once you're in the game, you've agreed to let someone else define the parameters." When this happens, of course, the question becomes "who defines the parameters. Who has created this environment, and what do they want you to believe?" As Herz bluntly (and wisely) reminds us, "If you're going to . . . fight a computer-mediated war — if you're going to play these games — it's a good idea to know who's making up the rules."

We have seen how the military, frequently in concert with corporate interests, has influenced educational institutions regarding the skills that are valued and taught, the way students are evaluated and sorted, and the methods and modes of instruction. That the military is now working so hard to recruit teenage gamers into its ranks illustrates one way in which the rise of the military-entertainment complex is determining which skills will be valued and nurtured in our children and how those skills will be applied — educationally, economically, and otherwise. The fact that technologically adept teenagers, by virtue of their popular entertainment practices, possess the very skills that the military now deems essential highlights not only the military's new approach to learning but also the contemporary confluence between war, entertainment, and education.

All but War Is Simulation, Redux

Every military and civilian official I've spoken with believes that the military's video game use will only expand — exponentially, most likely — in the coming years. The Pentagon's growing reliance on special operations forces, along with an increasing emphasis on drone attacks and online combat, will help to fuel this escalation. (For instance, U.S. Special Operations Command recently purchased NeuroTracker, a virtual-reality system that trains the brain for fast-paced, chaotic scenarios.) Military leaders believe that America's future wars will revolve around so-called hybrid scenarios. Much like Afghanistan and Iraq, these scenarios will involve the whole spectrum of military action — "from support to civil authorities," as Thom Shanker reports, "to training local security forces to counterinsurgency to counterterrorism raids to heavy combat." Virtual and video game–based training are considered essential to each of these specialized activities.

The military's game use will also continue to move beyond the realm of training, as with Virtual Iraq/Afghanistan and the games developed by WILL Interactive. For example, the navy is funding a project to hack used video game consoles as a means of gathering data, including chat-room information, about the consoles' previous owners in order to track potential terrorists and enemies around the globe. (For now, the project forbids going after American citizens.) DARPA and the navy have also started using games as crowdsourcing tools. The Office of Naval Research's MMOWGLI (Massive Multiplayer Online War Game Leveraging the Internet) challenged groups of online gamers to come up with innovative solutions for combating Somali pirates. The popular DARPA-funded online video game *Foldit* enables players to contribute to weighty scientific research. In one recent example, *Foldit* players deciphered, in a mere ten days, the protein structure of simian AIDS — a problem that had baffled scientists for more than fifteen years. *Foldit*'s success has inspired DARPA to invest millions of dollars in developing a game that will use crowdsourcing to help debug software code. (Unreliable software is a major drain on Pentagon funds.)

The applications for video games do not end there. In a radical move, the army is seeking to create avatars for every member of the force. *National Defense* reports that these virtual representations "would accompany service members throughout their training and allow them to see, through simulation, how their skills, or lack thereof, would play in life and death situations." These individualized avatars would be available twenty-four hours a day, seven days a week. According to Chester Kennedy, Lockheed Martin's vice president for engineering, global training, and logistics, "With avatar technology, you can take somebody today who experienced a new threat and have him role-play for those going into theater in real time . . . If you look at the training continuum, how many things can be satisfied by an [artificially intelligent] avatar today as opposed to two years or a year ago? We're continually dramatically improving."

Still, for all its efforts and interest in the video game realm, the military has taken an approach that has been haphazard at best. Certainly no identifiable strategic framework or set of priorities has guided the Pentagon. In several cases, the adoption of games has largely been the result of efforts by specific people within the Pentagon hierarchy or within the service branches — Casey Wardynski is an example — who have, through force of will or otherwise, maneuvered their way through a change-resistant bureaucracy in order to make their visions of effective digital learning a reality. (The ad hoc technological improvisations of soldiers in the field have been equally important in this regard.)

There are also aspects of warfare that are, of course, simply unpredictable. As one general told me, "Things are going to go wrong in the real world that you just can't predict. They're based on how humans behave, not on how a machine behaves." The "green on blue" attacks in Afghanistan, in which Afghan police and soldiers intentionally kill their NATO coalition counterparts, are an example of this. They provide a telling reminder of the limitations of technological solutions to real-world problems.

Overreliance on technological models and simulations was a hallmark of the Cold War years, when, Paul Edwards writes, American foreign policy became inextricably bound to "high-technology military strategy." This bond was reinforced within the military by an ethos of techno-rationality in which new technologies were seen as capable of overcoming the most difficult political and military circumstances — part of what led to such disaster in Vietnam. The use of video games represents both a symbolic and a practical update of this belief. As in the Cold War, the military discourse surrounding this technology often centers on human-machine integration, the centrality of the man-machine unit to systems-based military thinking. At its extreme, this discourse shades into what Edwards calls "fictions, fantasies, and ideologies" that include visions of battlefield oversight through centralized, instantaneous, computer-based command and control.

In many ways, the evolution of the military's game use has mirrored the Pentagon's painful learning process during the wars in Iraq and Afghanistan. At the risk of sounding flippant, I see the scenario as often going like this: Soldiers need to learn how to interact with Afghan warlords and Iraqi sheikhs? Give 'em a game. Soldiers need to learn basic communication skills in Arabic and Pashto? Give 'em a game. Soldiers need to learn how to manage IED attacks while running convoy operations? Give 'em a game. Thousands of veterans are suffering from PTSD? Give 'em a game. Video games can make it appear as if the military is effectively covering an issue that in fact needs greater attention and resources.

The Pentagon also remains challenged by issues of fidelity (the measure of realism) in simulation. Take the example of electronic warfare, in which the military wants radar on missiles to exhibit a high level of fidelity. This is possible in the air, where maybe only one hundred platforms are operating at a time. But as retired general Paul Kern told me, "When you take it down into the ground domain and you've got thousands of platforms operating, the simulations tend to get bogged down. How we scale all of that and how we manage that is one of the challenges we haven't figured out." Once logistics, with their millions of bits of information, are brought into the equation, the issue becomes even more complex. Everything from bullets to fuel to food has an impact on a real operation. How then, Kern asks, "can you get the same impact on you as a human being in decisions that you make without going to the level of fidelity of feeling a bullet enter your head?"

There are economic issues at play here, too. While the intellectual support for using video games is "huge," says Colonel Anthony Krogh, director of the army's National Simulation Center, that support doesn't always translate into resources. For example, of the $3 billion allotted annually to PEO STRI, only $20 million is dedicated to gaming per se (as opposed to more basic modeling and simulation, such as tank simulators). "That's pennies," Krogh told me. "It's budgetary dust compared to some other programs. And yet hundreds of thousands of soldiers are training with *VBS2*." (By way of comparison, a single Tomahawk missile costs $1 million. During the first twenty-four hours of U.S.-led bombing in Libya in 2011, at least 110 Tomahawks were fired.)

PEO STRI, for one, is striving to adapt to this situation. Colonel Franklin Espaillat, who oversees six product lines, including gaming, at PEO STRI, outlined the organization's mode of operation for me. "It's all about reuse for us," he said. "When we buy something, we reuse it across the product line wherever we can. With virtual training, the expense comes right at the beginning but diminishes after the initial capital investment." Because PEO STRI has a limited amount of research and development money, it relies on the corporate world to perform the bulk of its R&D work.

The constant tension between requirements and funding will probably define the military's game-related efforts for years to come, forcing people such as Krogh and Espaillat to continue balancing between the two. "We believe that with simulation we save lives," Espaillat says, "but our challenge is that we continually have to prove that it's good to the budget folks. We have to constantly show simulation training effectiveness so that we can keep these programs alive."

Frank DiGiovanni, the Department of Defense's director of training readiness and strategy — the senior Pentagon official in charge of training policy and oversight — acknowledges that the problem will continue with the military's current budget cuts. "The DoD is asking me to still have people who are ready for conflict, but to do it with much less funding," he told me. "Because I'm the training guy, I'll certainly make the case that investing in these technologies, you *will* get return on your investment."

Like many defense officials I've spoken with, DiGiovanni believes that the military should learn from commercial industry's development model. "Industry is very agile when it comes to fielding technology," he notes. "I would certainly like to see the department's training community be just as agile. There are reasons, of course, why we're maybe not as agile as we could be. But in the case of software- and technology-based things, not being agile sometimes allows technology to pass you by."

Cyberwar

Nowhere in the military will video games and related technologies play a more significant role than in the realm of cyberwar. This kind of warfare didn't exist a generation ago, and yet it may have a more pervasive and debilitating effect on countries at conflict than real-world combat. (At the very least, cyberwarfare will be an essential aspect of any coming major clash.) Brookings Institution defense expert Peter Singer refers to this new element of war as "battle-zone persuasion," in which the purpose "is not to blow up the enemy tank, but jam it, co-opt it, persuade it to do something that its owner doesn't want it to do. This is new in war." Seen in this regard, the Stuxnet and Flame computer viruses unleashed against Iran's nuclear program by the United States and Israel are notable not only for the actual damage they've inflicted but as harbingers of the future struggles between nation-states. As David Sanger reports, Stuxnet "appears to be the first time the United States has repeatedly used cyberweapons to cripple another country's infrastructure, achieving, with computer code, what until then could be accomplished only by bombing a country or sending in agents to plant explosives." With this threshold irrevocably crossed, America stands at the beginning of a new and uncertain era of conflict.

The Pentagon is slowly beginning to adjust itself to this changing landscape. According to the *Washington Post,* DARPA is reaching out to private enterprise, academia, and video gamers in an effort to upgrade its cyberwarfare capabilities. Among the aims of this new project, named Plan X, is "the creation of an advanced map that details [and 'continuously updates'] the entirety of cyberspace." Military commanders would use this map to pinpoint and disable enemy targets via computer code. Plan X also seeks "the creation of a new, robust operating system capable of both launching attacks and surviving counterattacks." Unlike the cyberattacks launched by the intelligence community, the military's attacks would focus on "achieving a physical effect," such as "shutting down or disrupting a computer." According to cyberexpert Herbert S. Lin of the National Academy of Sciences, "If they can do it, it's a really big deal . . . They're talking about being able to dominate the digital battlefield just like they do the traditional battlefield."

On a broader level, the Pentagon's new United States Cyber Command is charged with centralizing and coordinating the various cyberspace resources that exist throughout the military. Headed by National Security Agency director General Keith Alexander, Cyber Command is responsible both for protecting the Defense Department's information infrastructure and for developing new offensive and defensive cyberwar capacities across the services as a whole. In a time of across-the-board Pentagon budget cuts, cybersecurity is one of the few areas that will see an actual increase in its budget in the years ahead. What hasn't been decided is the extent to which Cyber Command will be able to respond to enemy attacks and how its duties will or will not overlap with those of the National Security Agency.

Critics of Cyber Command argue that its centralized, hierarchical structure is evidence of politics triumphing over effective policy. Cyberwarfare is by nature a decentralized, flattened, networked phenomenon; it therefore stands to reason that an organization like Cyber Command is the worst possible way of approaching it. The Pentagon could more sensibly follow the example of Russia and China and harness the power of independent hackers, whose expertise often far outshines that of government employees. Thus far, however, the Obama administration has taken the opposite tack, seeking to prosecute hackers wherever possible.

Every technological change implies a new way of waging battles. The rise of cyberwarfare precipitates what John Arquilla, a professor of defense analysis at the Naval Postgraduate School, terms the movement "from blitzkrieg to bitskrieg." One of the country's leading cyberwar experts, Arquilla believes the hallmark of the next century of conflict will be "the increasing ability to move from the virtual world to have effects in the physical world," and vice versa. This is where the questions arise. If cyberwar itself is a form of virtual reality, Arquilla asks, "how will it manifest concrete effects in the real world? Will taking out an enemy's power grid via a worm be considered an act of war?" By the same token, how will real-world military strategies and tactics manifest themselves in the virtual realm? The cross-pollination between the virtual and the real may well be, Arquilla says, the most important development in the realm of conflict over the next several decades.

To prepare for this future, the U.S. Air Force has started training officers to defend its electronic networks, pursue online hackers, and launch cyberattacks. In 2012 the air force's prestigious Weapons School, located at Nellis Air Force Base in Nevada, graduated its first class of cyberwarriors. According to Colonel Robert Garland, the Weapons School's commandant, "While cyber may not look or smell exactly like a fighter aircraft or a bomber aircraft, the relevancy in any potential conflict [today] is much the same . . . We have to be able to succeed against an enemy that wants to attack us in any way."

The handful of officers who made up the inaugural class of the Cyber Weapons Instructor Course at Nellis were drawn primarily from the 67th Network Warfare Wing and the 688th Information Operations Wing at San Antonio's Lackland Air Force Base. Before arriving at the Weapons School, they underwent three months of undergraduate cyber training at Keesler Air Force Base and an additional two months of intermediate network warfare training at Hurlbert Field, Florida.

Once at Nellis, the officers embarked on a grueling six-month curriculum of ten- to twelve-hour days.

The curriculum in Nellis's cyberweapons course is based on real-world scenarios. "We pick a region of the world where there would be increasing tensions and an adversary who doesn't want U.S. involvement in that area," says the lieutenant colonel in charge of the course. "And then we play out the computer side of that war in the virtual space and challenge our students with what they could expect to see from an adversary in that area." An enemy, for example, may try to penetrate the air force's systems in order to steal information about future operations; it may also deliver corrupt information into the system in order to disrupt those operations. The range of potential targets is wide: attackers may target an entire command system, or they may focus on a single airplane.

The students in the course are charged with building electronic defenses against this potential intrusion. "Aggressor squadron" teams at Nellis play as their opponents. "The Air Force aggressor acts as a hacker coming against us and we see how our defensive plans measured up," says Lieutenant Colonel Steven Lindquist, one of the eight initial students. The students learn offensive cyberattack capabilities as well, such as jamming enemy air and sea defenses, as Israel did during its 2007 airstrike on a Syrian nuclear reactor.

According to Lieutenant Colonel Bob Reeves, who directs the cyber course, the new cybercurriculum "is based on attack, exploit, and defense of the cyber domain." When graduates complete the course, they will go on to work for Cyber Command, where they will act as instructors and advisers for senior officials. They will be joined by members from other services, including the navy, which is overhauling cyber training at its Center for Information Dominance, where upward of 24,000 people train every year.

In some ways the rise of cyberwarfare would seem preferable to bloody on-the-ground combat, and yet the choice is not so simple. "If the bitskrieg world is one in which war is not as terrible," John Arquilla says, "then maybe war is not unthinkable." We see this issue playing out already with drone warfare in Pakistan, Yemen, and Somalia. If war is made less violent, less disruptive to society, will we be more inclined to wage it? "All but war is simulation," reads PEO STRI's former motto. But what happens when that equation changes — when war itself becomes simulation? We are several decades into the information age, but we have still not fully grappled with what that means for the American, and the global, way of war.

Nor have we fully confronted the fact that the United States has no monopoly on the innovative uses of virtual weapons and video games in war. A number of other countries, including China, South Korea, and the United Kingdom, have recently followed the Pentagon's lead and created their own cyberwarfare commands. Groups such as Hamas and Hezbollah, meanwhile, have produced their own first-person shooter military video games. In Hezbollah's *Special Force* series, players take on the role of combatants against the Israeli Defense Forces. The game box declares that the action "embodies objectively the defeat of the Israeli enemy and the heroic actions taken by heroes in Lebanon." Similarly, Syria's *Under Siege* features Palestinian fighters battling Israeli invading forces. And in Iran's *Special Operation 85: Hostage Rescue,* players must free a husband-and-wife team of nuclear scientists who have been kidnapped by the American military.

Perhaps most tellingly, the Chinese army recently announced the development of its own first-person shooter game, reportedly modeled on *America's Army.* Titled *Glorious Mission,* the game, like *America's Army,* requires players to complete basic training before advancing to online team combat. Unlike *America's Army*, after players complete training and combat, they move into a third stage, which re-creates what a Chinese news report calls "the fiery political atmosphere of camp life."

Another key difference between *Glorious Mission* and *America's Army:* the bad guys aren't unspecified Middle Eastern, Eastern European, or Central Asian terrorists. In *Glorious Mission,* there is one enemy only: the United States military.

NOTES

All quotations that are not specifically cited are drawn from the author's interviews with the persons quoted.

Introduction

4 *At the time, military psychologists:* Driskell and Olmstead, "Psychology and the Military."

5 *"Games are the future of learning":* Quoted in Singer, "Meet the Sims."

6 *Moreover, although commercial interests:* Li, "The Potential of America's Army the Video Game."

The entire game industry: Prensky, *Digital Game-Based Learning.*

This exchange has led scholars: The phrase "military-entertainment complex" comes from Bruce Sterling and is used by, among others, Timothy Lenoir and McKenzie Wark.

7 *According to the Defense Intelligence Agency:* Defense Intelligence Agency, "Informational Brief."

The 2012 Department of Defense budget: U.S. Department of Defense, *FY2012 Defense Budget,* A4, A11–12.

the research firm Frost & Sullivan predicts: National Training and Simulation Association, "Training 2015."

"*game-based training can be tailored:*" Singer, "Meet the Sims."

1. The Rise of the Military-Entertainment Complex

For this chapter, I would like to acknowledge the groundbreaking work of Ed Halter, Paul Edwards, Timothy Lenoir, Henry Lowood, J. C. Herz, Stephen Kline, Nick Dyer-Witheford, Greig de Peuter, Heather Chaplin, Aaron Ruby, and Sharon Ghamari-Tabrizi.

11 *"The technologies that shape our culture:"* Halter, *From Sun Tzu to Xbox*, 187.
While private industry may eventually have developed: Edwards, *Closed World*, 43.

12 *Though ENIAC wasn't completed:* Halter, *From Sun Tzu to Xbox*, 89.
"the proving ground for initial concepts": Edwards, *Closed World*, 43.

13 *In fact, computers were for many years:* Halter, *From Sun Tzu to Xbox*, 90.
The other major beneficiary: Kline, Dyer-Witheford, and de Peuter, *Digital Play*, 85.
the "military-industrial-academic complex": Ibid.

14 *Advanced computing systems, computer graphics:* Ibid., 99.
The game was invented in 1962: Brand, "Spacewar."
"the size of about three refrigerators": Halter, *From Sun Tzu to Xbox*, 74–75.
The PDP-1's manufacturer had shipped: Herz, *Joystick Nation*, 5.

15 *Russell's main influence in programming Spacewar!:* Brand, "Spacewar."
"Beams, rods, and lances of energy": Edward Smith, *Triplanetary*, quoted in ibid.
"picking a world": Ibid.

16 *"radical innovation":* Kline, Dyer-Witheford, and de Peuter, *Digital Play*, 87.
Within a year, the game had: Halter, *From Sun Tzu to Xbox*, 75.
By the mid-1960s: Herz, *Joystick Nation*, 7.

17 *Nuclear mobilization, ballistics, missilery, space defense:* Kline, Dyer-Witheford, and de Peuter, *Digital Play*, 85.
"were not created directly for military purposes": Halter, *From Sun Tzu to Xbox*, 82–83.
The military's specific interest in computer-based war gaming: Lenoir and Lowood, "Theaters of War," 6.

18 *Not only did Battlezone evoke a three-dimensional world:* Halter, *From Sun Tzu to Xbox*, 119.
"[Today's soldiers have] learned": Ibid., 136–37.

19 *Each of these systems cost:* Lenoir and Lowood, "Theaters of War," 10.
"Group interactions are the most complicated": Quoted in Hapgood, "Simnet."

20 *"William Gibson didn't invent cyberspace":* Ibid.
By January 1990, the first SIMNET *units:* Lenoir and Lowood, "Theaters of War," 16.

21 *"graphics and networking technology":* Ibid., 30.
Around the same time that Doom: Riddell, "Doom Goes to War."

22 *According to Barnett, Marines would plead:* Chaplin and Ruby, *Smartbomb*, 202.
"part-task training; mission rehearsal": Ghamari-Tabrizi, "Convergence of the Pentagon and Hollywood," 153.

23 *By contrast,* Marine Doom: Chaplin and Ruby, *Smartbomb*, 206.

24 *Two new declarations of military doctrine:* Ghamari-Tabrizi, "Convergence of the Pentagon and Hollywood," 155–56.

"As [Internal Look] got under way": Quoted in Lenoir and Lowood, "Theaters of War," 10.

30 *"sharing research results, coordinating research agendas"*: National Research Council, *Modeling and Simulation,* n.p.

2. Building the Classroom Arsenal: The Military's Influence on American Education

36 *"The emphasis on science and mathematics education"*: Noble, *Classroom Arsenal,* 14.

37 *the "vocationalism" movement:* Grubb and Lazerson, *Education Gospel,* vii.

Even today the military continues to boast: Zwick, *Fair Game?,* 2.

"a collateral investment": Brandt, "Drafting U.S. Literacy," 495.

The vast majority of these new recruits: Driskell and Olmstead, "Psychology and the Military," 46.

38 *For example, over a two-year period in World War I:* Ibid.

40 *The results of the Alpha led:* Resnick and Resnick, "The Nature of Literacy," 381.

gave rise to three "facts": Gould, *Mismeasure of Man,* 226–27.

41 *"produced a way of thinking"*: Sticht, *Military Experience and Workplace Literacy,* 21.

World War I also marked the first time: Driskell and Olmstead, "Psychology and the Military," 48.

By World War I, however: Sticht, *Military Experience and Workplace Literacy,* 16.

42 *Comprehension, not just decoding:* Duffy, "Literacy Instruction in the Military," 441.

"to develop arrested mentality": Egardner, "Adult Education in the Army," 258.

After World War I, the attention generated: Resnick and Resnick, "Nature of Literacy," 381.

"A certain division contains 5,000 artillery": Quoted in Zwick, *Fair Game?,* 2.

43 *Although intended as an examination of general learning ability:* Eitelberg, Laurence, Waters, and Perelman, *Screening for Service,* 15.

Between 1941 and 1945, the years of U.S. involvement: Lemann, *Big Test,* 54.

This in turn focused attention: Zwick, *Fair Game?,* 3.

Though unintended by its creators, the G.I. Bill: Kime and Anderson, "Contributions of the Military," 475.

44 *From this examination came a newfound:* Driskell and Olmstead, "Psychology and the Military," 48.

This program established permanently: Anderson, "Historical Profile," 59.

It also made education relevant: Kime and Anderson, "Contributions of the Military," 465.

Between 1941 and 1945, the minimum standards for enlistment: Brandt, "Drafting U.S. Literacy," 486–87.

Initially designed by the staff: Kime and Anderson, "Contributions of the Military," 468.

45 *Study by correspondence was viewed:* Anderson, "Historical Profile," 110.

"mediate technologies": Brandt, "Drafting U.S. Literacy," 485, 495.

46 *"Computers would probably have found":* Noble, *Classroom Arsenal,* 3.

"Which are the strong points": Quoted in ibid.

47 *Funded by the air force, army, and navy, PLATO was:* Fletcher, "Education and Training Technology in the Military."

For years PLATO was the world's: Noble, *Classroom Arsenal,* 98.

The SAGE system also pioneered: Ibid., 73–81.

48 *"The Knowledge Revolution":* Grubb and Lazerson, *Education Gospel,* 1–2.

the "major determinant": Quoted in Noble, *Classroom Arsenal,* 12.

"information theory, systems analysis, nuclear energy": Ibid., 191.

3. "Everybody Must Think": The Military's Post-9/11 Turn to Video Games

50 *A 2001 Army Science Board study laid out:* Army Science Board, "Manpower and Personnel."

51 *To do so would require "jointness":* Halter, *From Sun Tzu to Xbox,* xxi.

"decided that it needed to think less": Quoted in Silberman, "War Room."

52 *"the cognitive demands":* Ibid.

"speed, degree, and duration": Ibid.

53 *"where are the opposing forces":* Nieborg, "Changing the Rules of Engagement," 117.

54 *"Our military information tends to arrive":* Quoted in Li, "Potential of America's Army," 42–43.

57 *"People have been using simulations":* Quoted in Silberman, "War Room."

58 *"is critical to learning":* Quoted in Chaplin and Ruby, *Smartbomb,* 207.

59 *Macedonia says that the book was a major influence:* Harmon, "U.S. Military Embraces Video Games."

What military training tries to do: Halter, *From Sun Tzu to Xbox,* 198–99.

"Somebody throws a ball at you": Ibid.

"It does get really weird": Quoted in Chaplin and Ruby, *Smartbomb,* 201, 195.

a growing number of education scholars: Scholars who make this argument include James Gee, Kurt Squire, Constance Steinkuehler, Alice Daer, Henry Jenkins, Gail Hawisher, and Cindy Selfe.

61 *Gee claims that we must start thinking:* Gee, *What Video Games Have to Teach Us,* 13.

"If we compare what individuals do": Steinkuehler, "Cognition and Learning," 100–101.

62 *"a diverse population of soldiers"*: Singh and Dyer, "Computer Backgrounds of Soldiers," 22.
"all their lives": Silberman, "War Room."
"want to learn Army digital systems": Schaab and Dressel, *Training the Troops*, 4.
63 *"We have discovered that video game players"*: Quoted in Freeman, "Researchers Examine Video Gaming's Benefits."
"focus on relevant visual information": Ibid.
"At the core of these action video game-induced improvements": "Video Gaming Boosts Your Ability."
64 *"promote dynamic cognitive activity"*: Dawes and Dumbleton, "Computer Games in Education Project: Report."
In this way, sophisticated video games: Dipietro et al., "Towards a Framework."
"develop the situated understandings": Jenkins et al., "Confronting the Challenges," 4.
"a set of cultural competencies": Ibid.
65 *"Play—the capacity to experiment"*: Ibid.
"import skill and drill exercises": Rice, "Assessing Higher Order Thinking in Video Games."
66 *"'hypothesize, probe the world'"*: Gee, *What Video Games Have*, 216.
67 *"is often built around the 'content fetish'"*: Gee, "Learning by Design."
68 *"Live field training is very expensive"*: Quoted in Minton, "Software."
69 *"instead of tornadoes, earthquakes, and Godzilla"*: Mockenhaupt, "SimCity Baghdad."

4. *America's Army:* The Game

75 *"30 percent of all Americans"*: Singer, "Meet the Sims."
79 *"Our [selling] strategy must be based"*: McHugh, "Army Rolls Out Big Guns," 22.
"an employer of last resort": Ibid.
83 *this approach ignores the facts:* Tversky and Kahneman, "Availability," 207.
86 *"to build a game-oriented, virtual reality-based"*: Keith Hattes e-mail to Michael Zyda et al., "Research Initiative for Army Recruiting," September 3, 1999.
88 *"'moral rot' that had become symbolically associated"*: Bartlett and Lutz, "Disciplining Social Difference," 122.
Attributing a quasi-religious role: Edwards, *Closed World*, 10.
90 *In addition to Mike Capps:* Davis et al., "Making America's Army."
100 *"The game's training module cost"*: Singer, "Meet the Sims."

5. All but War Is Simulation

107 *"the training [positively] impacted"*: Ratwani, Orvis, and Knerr, "Game-Based Training Effectiveness Evaluation."

108 *"You can't simulate the dust, dirt"*: Martin and Lin, "Keyboards First. Then Grenades."

6. WILL Interactive and the Military's Serious Games

116 *"These games have an explicit"*: Abt, *Serious Games*, 9.
"a mental contest, played with a computer": Zyda, "From Visual Simulation to Virtual Reality to Games," 26.
125 *"I was injured in 2006"*: Quoted in Dao, "Acting Out War's Inner Wounds."

7. The Aftermath: Medical Virtual Reality and the Treatment of Trauma

136 *"looks like it has met a boot"*: Halpern, "Virtual Iraq."
142 *According to the* Journal of CyberTherapy and Rehabilitation: Wiederhold and Wiederhold, "Virtual Reality."
An issue of Studies in Health Technology: Parsons et al., "Neurocognitive and Psychophysiological Analysis."
143 *Most recently,* Military Medicine: McLay et al., "Development and Testing of Virtual Reality Exposure Therapy."
A recent issue of Psychiatric Services: Kramer et al., "Clinician Perceptions."
"is tough. It's tough": WILL Interactive brochure, n.d.
145 *A recent Mental Health Advisory Team study:* Casey, "Comprehensive Soldier Fitness."
146 *"to find any evidence of a well-coordinated"*: American Psychological Association, "Psychological Needs."
150 *"a structured, long-term assessment and development program"*: Comprehensive Soldier Fitness, http://csf2.army.mil/about.html.

8. Conclusion: *America's Army* Invades Our Classrooms

154 *"promote student interest in the engineering and technical fields"*: Welburn, "Upcoming Events."
156 *"stimulate, and sustain, dialogue"*: Ibid.
"have the opportunity to test": National Association of State Boards of Education, "Announcement," August 25, 2008. www.nasbe.org/index.php/upcomingevents/details/29-us-army.
157 *"reaffirming and strengthening America's role"*: White House, "Educate to Innovate," www.whitehouse.gov/issues/education/k-12/educate-innovate.
"global competitiveness": Burke and McNeill, "'Educate to Innovate.'"
158 *"a multi-year competition whose goal"*: stemchallenge.org.
159 *The* New York Times Magazine *reports that a growing number:* Corbett, "Learning by Playing."

"the big challenge isn't": Quoted in Silberman, "War Room."

160 *"been driven . . . by the real and pressing needs"*: Noble, *Classroom Arsenal*, 1.

"the purposes and the 'products' of education": Ibid., 6–7.

"the particular goals of those parties": Ibid., 4.

"politics by other means": Starr, "Seductions of Sim."

"The one thing . . . political simulations": Herz, *Joystick Nation*, 223.

"social contract": Ibid.

161 *U.S. Special Operations Command recently purchased:* Drummond, "Commandos Now Play Digital Brain Games."

162 *"from support to civil authorities"*: Shanker, "Army Will Reshape Training."

The popular DARPA-funded online video game Foldit: Lim, "Agencies Get Down to Business."

"would accompany service members": Beidel, "Avatars Invade Military Training Systems."

163 *"high-technology military strategy"*: Edwards, *Closed World*, 7–8, 15.

166 *"battle-zone persuasion"*: Quoted in Peck, "Since When Does Brookings Make Video Games?"

"appears to be the first time": Sanger, "Obama Order."

167 *"the creation of an advanced map"*: Nakashima, "With Plan X."

168 *"While cyber may not look or smell"*: Quoted in Barnes, "Pentagon Digs In on Cyberwar Front."

169 *"We pick a region of the world"*: Quoted in Toplikar, "Teaching the Shadowy Art of Cyber War."

"The Air Force aggressor acts as a hacker": Quoted in Barnes, "Pentagon Digs In on Cyberwar Front."

"is based on attack, exploit, and defense": Ibid.

170 *"the fiery political atmosphere of camp life"*: Axe, "Gamers Target U.S. Troops."

BIBLIOGRAPHY

Abt, Clark. *Serious Games.* New York: Viking, 1970.

American Psychological Association. Presidential Task Force on Military Deployment Services for Youth, Families and Service Members. "The Psychological Needs of U.S. Military Service Members and Their Families: A Preliminary Report." 2007. www.apa.org/releases. Accessed April 1, 2013.

Anderson, C. L. "Educating the United States Army." In Michael D. Stephens, ed., *The Educating of Armies,* pp. 39–74. New York: St. Martin's, 1989.

———. "Historical Profile of Adult Basic Education Programs in the United States Army." PhD dissertation, 1986, Teachers College, Columbia University.

Army Science Board. "Manpower and Personnel for Soldier Systems in the Objective Force." Washington, DC: Department of the Army, 2001.

Au, Wayne. *Unequal by Design: High-Stakes Testing and the Standardization of Inequality.* New York: Routledge, 2008.

Axe, David. "Gamers Target U.S. Troops in Chinese Military 'Shooter.'" *Danger Room,* May 17, 2011. www.wired.com/dangerroom/2011/05/gamers-target-u-s-troops-in-chinese-military-shooter/. Accessed March 17, 2013.

Barnes, Julian E. "Pentagon Digs In on Cyberwar Front." *Wall Street Journal,* July 6, 2012. http//online.wsj.com/article/SB10001424052702303684004577508850690121634.html. Accessed November 4, 2012.

Bartlett, Lesley, and Elizabeth Lutz. "Disciplining Social Difference: Some Cultural Politics of Military Training in Public High Schools." *Urban Review* 30, no. 2 (1998): 119–36.

Beidel, Eric. "Avatars Invade Military Training Systems." *National Defense Magazine,* February 2012. www.nationaldefensemagazine.org/archive/2012/February/Pages/AvatarsInvadeMilitaryTrainingSystems.aspx. Accessed March 13, 2013.

Binkin, M., M. Eitelberg, A. Schexnider, and M. Smith. *Blacks and the Military.* Washington, DC: Brookings Institution, 1982.

Braddock, J., and R. Chatham. "Defense Science Board Task Force on Training for Future Conflicts Final Report." Washington, DC: Office of the Undersecretary of Defense for Acquisition, Training, and Logistics, 2003.

Brand, Stewart. "Spacewar: Fanatic Life and Symbolic Death Among the Computer Bums." *Rolling Stone,* December 1972. www.wheels.org/spacewar/store/rolling-stone.html. Accessed March 16, 2013.

Brandt, Deborah. "Drafting U.S. Literacy." *College English* 66, no. 5 (2004): 485–502.

Bryant, B. "Friends and Colleagues." Letter to members of the National Association of State Boards of Education, n.d. www.susanohanian.org/outrage_fetch.php?id=510.

Burke, Lindsey, and Jena Baker McNeill. "'Educate to Innovate': How the Obama Plan for STEM Education Falls Short." Heritage Foundation, January 5, 2011. www.heritage.org/research/reports/2011/01/educate-to-innovate-how-the-obama-plan-for-stem-education-falls-short. Accessed April 1, 2013.

Card, Orson Scott. *Ender's Game.* New York: Starscape, 1985.

Casey, George W., Jr. "Comprehensive Soldier Fitness: A Vision for Psychological Resilience in the U.S. Army." *American Psychologist* 66, no. 1 (January 2011): 1–3.

Chaplin, Heather, and Aaron Ruby. *Smartbomb.* Chapel Hill, NC: Algonquin, 2005.

Clark, Harold F., and Harold S. Sloan. *Classrooms in the Military: An Account of Education in the Armed Forces of the United States.* New York: Teachers College Press, 1964.

Corbett, Sara. "Learning by Playing: Video Games in the Classroom." *New York Times Magazine,* September 15, 2010. www.nytimes.com/2010/09/19/magazine/19video-t.html?pagewanted=all$_r=0. Accessed March 17, 2013.

Dao, James. "Acting Out War's Inner Wounds." *New York Times,* January 2, 2012, A1.

Davis, Margaret, ed. *America's Army PC Game Vision and Realization.* San Francisco: U.S. Army/MOVES Institute, 2004.

Dawes, L., and T. Dumbleton. "Computer Games in Education Project: Report." Coventry, Eng.: Becta, 2001. http://tna.europarchive.org/20080509164701/partners.becta.org.uk/index.php?section=rh&rid=13595. Accessed March 29, 2013.

Defense Intelligence Agency. "Informational Brief: Virtual Worlds — Exploration and Application to Undersea Warfare." www.public.navy.mil/subfor/underseawarfare magazine/issues/archives/issue_45/virtual_worlds.html. Accessed May 26, 2013.

Dipietro, Meredith, Richard Ferdig, Jeff Boyer, and Erik Black. "Towards a Framework for Understanding Electronic Educational Gaming." *Journal of Educational Multimedia and Hypermedia* 16, no. 3 (2007): 225–48.

Driskell, James E., and Beckett Olmstead. "Psychology and the Military: Research Applications and Trends." *American Psychologist* 44, no. 1 (1989): 43–54.

Drummond, Katie. "Commandos Now Play Digital Brain Games as War Prep." *Danger Room,* May 8, 2012. www.wired.com/dangerroom/2012/05/socom-neurotracker/. Accessed March 29, 2013.

Duffy, Thomas. "Literacy Instruction in the Military." *Armed Forces & Society* 11, no. 3 (1985): 437–67.

Edwards, Paul. *The Closed World: Computers and the Politics of Discourse in Cold War America.* Cambridge, MA: MIT Press, 1996.

Egardner, Zacheus Tom. "Adult Education in the Army." *School Review* 30, no. 4 (1922): 255–67.

Eitelberg, M., J. Laurence, B. Waters, and L. Perelman. *Screening for Service: Aptitude and Education Criteria for Military Entry.* Washington, DC: Office of Assistant Secretary of Defense, Human Resources Research Organization, 1984.

Fletcher, J. D. "Education and Training Technology in the Military." *Science* 323 (2009): 72–75.

Freeman, Bob. "Researchers Examine Video Gaming's Benefits." www.defense.gov/news/newsarticle.aspx?id=57695. Accessed July 1, 2012.

Gee, James Paul. "Good Video Games and Good Learning." www.academiccolab.org/resources/documents/Good_Learning.pdf. Accessed March 16, 2013.

———. *Good Video Games and Good Learning: Collected Essays on Video Games, Learning, and Literacy.* New York: Peter Lang, 2007.

———. "Learning by Design: Good Video Games as Learning Machines." *E-Learning and Digital Media* 2, no. 1 (2005).

———. *What Video Games Have to Teach Us About Learning and Literacy.* New York: Palgrave Macmillan, 2003.

Ghamari-Tabrizi, Sharon. "The Convergence of the Pentagon and Hollywood." In Lauren Rabinovitz and Abraham Geil, eds., *Memory Bytes,* pp. 150–74. Durham, NC: Duke University Press, 2004.

Ginzberg, Eli, and Douglas W. Bray. *The Uneducated.* New York: Columbia University Press, 1953.

Goldberg, Samuel. *Army Training of Illiterates in World War II.* New York: Teachers College Press, 1951.

Gould, Stephen J. *The Mismeasure of Man.* New York: W. W. Norton, 1996.

Grubb, W. Norton, and Marvin Lazerson. *The Education Gospel: The Economic Power of Schooling.* Cambridge, MA: Harvard University Press, 2004.

Halpern, Sue. "Virtual Iraq." *The New Yorker,* May 19, 2008. www.newyorker.com/reporting/2008/05/19/080519fa_fact_halpern. Accessed March 17, 2013.

Halter, Ed. *From Sun Tzu to Xbox: War and Video Games.* New York: Thunder's Mouth, 2006.

Hapgood, Fred. "Simnet." *Wired,* May 2004. www.wired.com/wired/archives/5.04/ff_simnet.html. Accessed March 17, 2013.

Harmon, Amy. "U.S. Military Embraces Video Games." *Taipei Times,* April 6, 2003, 11.

Herz, J. C. *Joystick Nation: How Videogames Ate Our Quarters, Won Our Hearts, and Rewired Our Minds.* New York: Little, Brown, 1997.

Herz, J. C., and Michael Macedonia. "Computer Games and the Military: Two Views." *Defense Horizons* 11 (2002): 1–8.

Jenkins, Henry, et al. "Confronting the Challenges of Participatory Culture: Media Education for the 21st Century." Occasional Paper. Chicago: MacArthur Foundation, 2006.

Kime, Steve F., and Clinton L. Anderson. "Contributions of the Military to Adult and Continuing Education." In Arthur L. Wilson and Elisabeth R. Hayes, eds., *Handbook of Adult and Continuing Education,* pp. 464–79. San Francisco: Jossey, 2000.

Kline, Stephen, Nick Dyer-Witheford, and Greig de Peuter. *Digital Play.* Montreal: McGill-Queen's University Press, 2003.

Kramer, Teresa, et al. "Clinician Perceptions of Virtual Reality to Assess and Treat Returning Veterans." *Psychiatric Services* 61, no. 11 (November 2010): 1153–56.

Lemann, Nicholas. *The Big Test: The Secret History of the American Meritocracy.* New York: Farrar, Straus, and Giroux, 1999.

Lenoir, Tim, and Henry Lowood. "Theaters of War: The Military-Entertainment Complex." In Jan Lazardzig, Helmar Schramm, and Ludger Schwarte, eds., *Kunstkammer, Laboratorium, Bühne — Schauplätze des Wissens im 17. Jahrhundert,* pp. 432–64. Berlin: de Gruyter, 2003. Citations from www.stanford.edu/dept/HPS/TimLenoir/Publications/Lenoir-Lowood_TheatersOfWar.pdf.

Lenoir, Timothy. "All but War Is Simulation: The Military-Entertainment Complex." *Configurations* 8 (2000): 238–335.

Li, Zhan. "The Potential of America's Army the Video Game as Civilian-Military Public Sphere." Master's thesis, 2004, Massachusetts Institute of Technology. http://dspace.mit.edu/handle/1721.1/39162. Accessed February 12, 2013.

Lim, Dawn. "Agencies Get Down to Business with Computer Games." Nextgov.com, April 2, 2012. www.nextgov.com/defense/2012/04/agencies-get-down-to-business-with-computer-games/50942/. Accessed March 17, 2013.

Macedonia, Michael. "Games, Simulation, and the Military Education Dilemma." 2001. http://net.educause.edu/ir/library/pdf/ffpiu018.pdf. Accessed April 1, 2013.

Martin, Andrew, and Thomas Lin, "Keyboards First. Then Grenades." *New York Times,* May 1, 2011. wwwnytimes.com/2011/05/02/technology/02wargames.html?pagewanted=all$_4=0. Accessed March 13, 2013.

McHugh, Jane. "Army Rolls Out Big Guns to Boost Recruiting." *Army Times,* August 30, 1999.

McLay, Robert, et al. "Development and Testing of Virtual Reality Exposure Therapy for Post-Traumatic Stress Disorder in Active Duty Service Members Who Served in Iraq and Afghanistan." *Military Medicine* 177, no. 6 (2012): 635–42.

Mead, Corey. "America's Army Invades Our Classrooms." *Rethinking Schools* 23, no. 4 (2009): 48–49.

Minton, Eric. "Software." *Today's Officer Magazine,* Summer 2005. www.moaa.org/Todays Officer/Magazine/Summer 2005/software.asp. Accessed July 18, 2005.

Mockenhaupt, Brian. "SimCity Baghdad." *The Atlantic,* January/February 2010. www.theatlantic.com/magazine/archive/2010/01/simcity_baghdad/307830. Accessed April 1, 2013.

Nakashima, Ellen. "With Plan X, Pentagon Seeks to Spread U.S. Military Might to Cyberspace." *Washington Post,* May 30, 2012. articles.washingtonpost.com/2012-05-30/world/35458424_1_cyberwarfare-cyberspace-pentagon-agency. Accessed November 4, 2012.

National Association of State Boards of Education. "State Education Leaders Join with U.S. Army." Press release, September 16, 2008.

National Research Council. Committee on Modeling and Simulation. "Modeling and Simulation: Linking Entertainment and Defense." Washington, DC: National Academies Press, 1997.

National Training and Simulation Association. "Training 2015." www.trainingsystems.org/publications/AirForce.pdf. Accessed June 11, 2012.

Nieborg, David B. "Changing the Rules of Engagement: Tapping Into the Popular Culture of *America's Army,* the Official U.S. Army Computer Game." Master's thesis, 2005, Utrecht University.

Noble, Douglas D. *The Classroom Arsenal: Military Research, Information Technology and Public Education.* London: Falmer, 1991.

Parsons, T. D., et al. "Neurocognitive and Psychophysiological Analysis of Human Performance Within Virtual Reality Environments." *Studies in Health Technology and Informatics* 142 (2009): 556–61.

Peck, Michael. "Since When Does Brookings Make Video Games?" foreignpolicy.com, May 8, 2012. www.foreignpolicy.com/articles/2012/05/08/call_of_duty_black_ops_interview. Accessed August 4, 2012.

Prensky, Mark. *Digital Game-Based Learning.* New York: McGraw-Hill, 2001.

Ratwani, Krista, Kara Orvis, and Bruce Knerr. "Game-Based Training Effectiveness Evaluation in an Operational Setting." Arlington, VA: U.S. Army Institute for the Behavioral and Social Sciences, 2010.

Resnick, Daniel P., and Lauren B. Resnick. "The Nature of Literacy: An Historical Exploration." *Harvard Educational Review* 47 (1977): 370–85.

Rice, John W. "Assessing Higher Order Thinking in Video Games." *Journal of Technology and Teacher Education,* 15:1 (2007): 87–100.

Riddell, Rob. "Doom Goes to War." *Wired* 5, no. 4 (April 1997). www.wired/com/wired/archive/5.04/ff_doom.html. Accessed March 17, 2013.

Sanger, David. "Obama Order Sped Up Wave of Cyberattacks Against Iran." *New York Times,* June 1, 2012, A1.

Schaab, Brooke B., and J. D. Dressel. *Training the Troops: What Today's Soldiers Tell Us About Training for Information-Age Digital Competency.* Alexandria, VA: U.S. Army Research Institute for the Behavioral and Social Sciences, 2003.

Shanker, Thom. "Army Will Reshape Training, with Lessons from Special Forces." *New York Times,* May 2, 2012, A22.

Silberman, Steve. "The War Room." *Wired* 12, no. 9 (September 2004). www.wired.com/wired/archive/12.09/warroom.html. Accessed March 14, 2013.

Singer, Peter W. "Meet the Sims . . . and Shoot Them." *Foreign Policy,* March/April 2010. www.foreignpolicy.com/articles/2010/02/22/meet_the_sims_and_shoot_them. Accessed March 13, 2013.

Singh, Harnam, and Jean L. Dyer. "The Computer Backgrounds of Soldiers in Army Units: FY01." Alexandria, VA: U.S. Army Institute for the Behavioral and Social Sciences, 2002.

Smith, Merritt Roe, ed. *Military Enterprise and Technological Change.* Cambridge, MA: MIT Press, 1985.

Starr, Paul. "Seductions of Sim: Policy as a Simulation Game." *American Prospect* 17 (1994): 19–29.

Steinkuehler, Constance. "Cognition and Learning in Massively Multiplayer Online Games: A Critical Approach." PhD dissertation, 2005. University of Wisconsin–Madison.

Sterling, Bruce. "War Is Virtual Hell." *Wired* 1, no. 1 (1993).

Sticht, Thomas G. *The Military Experience and Workplace Literacy: A Review and Synthesis for Policy and Practice.* National Center on Adult Literacy Technical Report. Philadelphia: National Center on Adult Literacy, 1995.

———. "The Rise of the Adult Education and Literacy System in the United States: 1600–2000." Jessup, MD: Office of Educational Improvement and Research, 2002.

Sticht, Thomas G., W. B. Armstrong, D. T. Hickey, and J. S. Caylor. *Cast-Off Youth: Policy and Training Methods from the Military Experience.* New York: Praeger, 1987.

Toplikar, Dave. "Teaching the Shadowy Art of Cyber War." *Las Vegas Sun,* July 6, 2012. www.lasvegassun.com/news/2012/jul/06/learning-art-cyber-war/. Accessed November 6, 2012.

Tversky, Amos, and Daniel Kahneman. "Availability: A Heuristic for Judging Frequency and Probability." *Cognitive Psychology* 5 (1973): 207–32.

U.S. Department of Defense. Office of the Undersecretary of Defense (Comptroller). *FY2012 Defense Budget—Research, Development, Test, and Evaluation Programs,* February 2011. http://comptroller.defense.gov/defbudget/fy2012/fy2012_r1.pdf.

"Video Gaming Boosts Your Ability to Concentrate in a Crisis." *Daily Mail Online,* November 19, 2010. www.dailymail.co.uk/health/article_1330807/Video-gaming_BOOSTS_ability_concentrate_Crisis.html. Accessed July 1, 2012.

Wark, McKenzie. *A Hacker Manifesto.* Cambridge, MA: Harvard University Press, 2004.

Welburn, B. "Upcoming Events: US Army and American Public Education: Building Strong Futures Together." 2009. www.nasbe.org/index.php/upcoming-events/details/29-us-army.

Welch, Nancy. "'This Video Game We Call War': Multimodal Recruitment in America's Army Game." *Reflections* 7, no. 3 (2008): 162–91.

White, Bruce. "ABC's for the American Enlisted Man: The Army Post School System, 1866–1898." *History of Education Quarterly* 8, no. 4 (1968): 479–96.

Wiederhold, Brenda K., and Mark D. Wiederhold. "Virtual Reality for Posttraumatic Stress Disorder and Stress Inoculation Training." *Journal of CyberTherapy and Rehabilitation* 1, no. 1 (Spring 2008): 23–35.

Yoakum, Clarence S., and Robert M. Yerkes. *Army Mental Tests.* New York: Holt, 1920.

Zwick, Rebecca. *Fair Game?: The Use of Standardized Admissions Tests in Higher Education.* New York: Routledge, 2002.

Zyda, Michael. "From Visual Simulation to Virtual Reality to Games." *IEEE Computer* 38, no. 9 (2005): 25–32.

ACKNOWLEDGMENTS

Deepest thanks to my agent and friend E. J. McCarthy, who not only responded to my initial e-mail query within three minutes but since then has been an unending source of wisdom, support, insight, good humor, and all-around positive energy. E. J., you are, quite simply, the best.

A big thank-you to Eamon Dolan, editor extraordinaire, for taking a chance on a completely unknown writer, for conceptual and structural brilliance, for always asking the toughest — and most necessary — questions, for patience in the face of my repeated attempts to write a completely different book from the one I'd proposed, and for being the kind of rigorous, committed, always-available editor that I'd assumed no longer existed.

Thank you to Tracy Walsh for her impeccable research and transcription skills. In addition to many other things, Tracy provided drafts of the material on *VBS2,* Virtual Iraq/Afghanistan, and *America's Army: True Soldiers.* Tracy has ahead of her a long and brilliant career as a writer and scholar.

My thanks as well to the dozens of officials, military and civilian, who spoke with me for my book and to the many hardworking PA officers and assistants who made those interviews happen.

At Baruch College, I offer heartfelt thanks to, among others, Cheryl

Smith, Frank Cioffi, Shelly Eversley, Sean O'Toole, Gina Parmar, Claudye James, and the PSC-CUNY Foundation. Thanks also to my wonderful students, whose intelligence, curiosity, and humor never cease to impress me.

Thank you to the CUNY-wide Composition and Rhetoric Group, especially Amy Wan, Tim McCormack, Mark McBeth, Leigh Jones, and Ericka Kaufman. At the Graduate Center, I thank Sondra Perl and Ira Shor for their support.

In Madison (and now beyond), I thank Wayne Au, Scot Barnett, Mike Bernard-Donals, Ross Collin, Alice Daer, Rasha Diab, Stephanie Fiorelli, Mary Fiorenza, David Fleming, Melanie Hoftyzer, Brad Hughes, Rik Hunter, Adam Koehler, Annie Massa-MacLeod, Martin Nystrand, Shifra Sharlin, Mira Shimabukuro, Kurt Squire, Constance Steinkuehler, Annette Vee, and Kate Vieira. My sincere apologies to anyone whom I may have overlooked.

Deborah Brandt was the ideal adviser, and the rigor, depth, and quality of her advice, criticism, and support (not to mention her own work) are without parallel. I cannot thank her enough for all the brilliant counsel and assistance she has given me.

Thanks to Scott Adkins and the Brooklyn Writers Space for the ideal setting and community in which to write.

True compadres: Karla Pazzi, Alex Pearcy, Jesse Seldess, Matt Vandre, Rob Voedisch. All of my friends in Brooklyn and beyond. Mom, Dad, and Ken, your lifelong love and encouragement have sustained me.

I have the world's best parents, brother, sister-in-law, niece, cousins, uncles, and aunts, and I love them and thank them for everything.

Laura, I cannot put into words how much your love and support mean to me. You and Caleb are the two best things that have ever happened to me, and I can't wait to spend a lifetime with both of you. It is for you, first and foremost, that this book is written.

INDEX